Titanium Alloys for Biomedical Implants and Devices

Titanium Alloys for Biomedical Implants and Devices

Editors

Hooyar Attar
Damon Kent

MDPI • Basel • Beijing • Wuhan • Barcelona • Belgrade • Manchester • Tokyo • Cluj • Tianjin

Editors

Hooyar Attar
The University of Queensland
Australia

Damon Kent
University of the Sunshine Coast
Australia
University of Queensland
Australia

Editorial Office
MDPI
St. Alban-Anlage 66
4052 Basel, Switzerland

This is a reprint of articles from the Special Issue published online in the open access journal *Metals* (ISSN 2075-4701) (available at: https://www.mdpi.com/journal/metals/special_issues/titanium_biomedical_application).

For citation purposes, cite each article independently as indicated on the article page online and as indicated below:

LastName, A.A.; LastName, B.B.; LastName, C.C. Article Title. *Journal Name* **Year**, *Volume Number*, Page Range.

ISBN 978-3-0365-0002-7 (Hbk)
ISBN 978-3-0365-0003-4 (PDF)

Contents

About the Editors

Hooyar Attar received his Ph.D. in Metal Additive Manufacturing in 2015. After completing Ph.D., Dr. Attar worked (2015–2016) as a Lecturer at Edith Cowan University for teaching and research activities. He then joined the University of Queensland (2016–2019) as a Postdoctoral Research Fellow and has worked on metal 3D printing and conventional manufacturing of high-performance metallic alloys and composites under onetime the largest Additive Manufacturing Hub in Australia (Industrial Transformation Research Hub (ITRH) for Additive Manufacturing). He is currently working as an R&D Researcher at Bradken and is also an Honorary Research Fellow at the University of Queensland.

Damon Kent is an Associate Professor of Engineering Sciences at the University of the Sunshine Coast (Australia) and is also Honorary Senior Fellow at The University of Queensland (Australia). He currently servs as Project Leader on the Australian Research Council (ARC) Research Hub for the Advanced Manufacturing of Medical Devices. His research is broadly focused on new materials and processing technologies for biomedical and aerospace applications.

Preface to "Titanium Alloys for Biomedical Implants and Devices"

The development of advanced titanium alloys and processing methods toward fabrication of biomedical implants and devices is an active field of research. Metallic biomaterials offer greater strength and toughness in comparison to polymers and ceramics and among them titanium-based alloys provide significant advantages, including higher strength to weight, excellent biocompatibility and stiffness which more closely resembles human hard tissue. Over the past few decades, a large amount of research has focused on various aspects of titanium alloys leading to the design, production, and commercialization of a series of alloys specifically tailored to biomedical applications. Despite the outstanding properties afforded by current titanium alloys, there is a need to continue to enhance their performance through developing further understanding of important aspects of their processing and structure. This is necessary to enhance the performance and reliability of titanium implants and devices and, consequently, improve patient health outcomes.

This Special Issue highlights challenges and recent advances in the application of titanium alloys for biomedical implants and devices. Featuring prominently, additive manufacturing to process titanium alloys is the focus of the first four articles. Currently, research in this area is burgeoning along with increasing clinical uptake, offering the promise of complex and customized implant and device designs. Issues highlighted in these articles include application of the technologies to fabricate complex porous scaffolds and assessing their clinical performance, and the impacts on microstructure and, consequently, performance obtained through different additive manufacturing technologies. Despite these advances, research is also ongoing to advance more conventional approaches to processing such as highlighted in articles on the improving fatigue performance, machinability and formability of titanium alloys. Finally, the issue highlights the need to improve long-term clinical performance of devices and implants including through modeling to assess biomechanical performance and understanding of corrosion behaviors. We believe the articles provide a current snapshot of the recent advances, as well as ongoing challenges in this exciting field. We thank all authors for their contributions to this special issue.

Hooyar Attar, Damon Kent
Editors

Article

Mechanical Properties and In Vitro Behavior of Additively Manufactured and Functionally Graded Ti6Al4V Porous Scaffolds

Ezgi Onal [1,*], Jessica E. Frith [1], Marten Jurg [1], Xinhua Wu [1,2] and Andrey Molotnikov [1,*]

[1] Department of Materials Science and Engineering, Monash University, Clayton, VIC 3800, Australia; jessica.frith@monash.edu (J.E.F.); marten.jurg@monash.edu (M.J.); xinhua.wu@monash.edu (X.W.)
[2] Monash Centre for Additive Manufacturing, Monash University, Clayton, VIC 3800, Australia
* Correspondence: ezgi.onal@monash.edu (E.O.); andrey.molotnikov@monash.edu (A.M.); Tel.: +61-3-9905-0398 (E.O.); +61-3-9905-9996 (A.M.)

Received: 15 February 2018; Accepted: 19 March 2018; Published: 21 March 2018

Abstract: Functionally graded lattice structures produced by additive manufacturing are promising for bone tissue engineering. Spatial variations in their porosity are reported to vary the stiffness and make it comparable to cortical or trabecular bone. However, the interplay between the mechanical properties and biological response of functionally graded lattices is less clear. Here we show that by designing continuous gradient structures and studying their mechanical and biological properties simultaneously, orthopedic implant design can be improved and guidelines can be established. Our continuous gradient structures were generated by gradually changing the strut diameter of a body centered cubic (BCC) unit cell. This approach enables a smooth transition between unit cell layers and minimizes the effect of stress discontinuity within the scaffold. Scaffolds were fabricated using selective laser melting (SLM) and underwent mechanical and in vitro biological testing. Our results indicate that optimal gradient structures should possess small pores in their core (~900 μm) to increase their mechanical strength whilst large pores (~1100 μm) should be utilized in their outer surface to enhance cell penetration and proliferation. We suggest this approach could be widely used in the design of orthopedic implants to maximize both the mechanical and biological properties of the implant.

Keywords: selective laser melting; gradient structure; porous biomaterial; Ti6Al4V; mechanical properties; osteoblast

1. Introduction

Recent advances in additive manufacturing have revealed new possibilities for the design of the next generation of metallic biomedical implants based on lattice structures. Generally, a bone scaffold should possess four essential characteristics [1,2]: (i) biocompatibility; (ii) mechanical properties matching those of the host tissue; (iii) an interconnected porous structure for cell migration and proliferation and nutrient–waste transportation; (iv) suitable surface characteristics for cell attachment. Traditionally, large bone defects are treated with metallic implants. Several metals have been shown to fulfil the requirement of biocompatibility including cobalt-based alloys, stainless steels as well as titanium alloys [3]. However, metallic implants possess higher elastic moduli than bone, e.g., Ti6Al4V and 316 L stainless steel have a modulus of around 110 GPa and 210 GPa [4] respectively, whereas the modulus of cortical bone is in the range of 3–20 GPa [5]. The mismatch between the modulus of the bone and the implant results in the failure of the implant in the long-term due to the stress-shielding problem [6]. Therefore, matching the mechanical properties of the implant to the host bone and simultaneously providing the implant with biological performance remains a challenge. One potential strategy is to create porous metallic scaffolds where the porosity, the pore size and shape are optimized collectively to reduce the modulus while maintaining the strength of the scaffold [7].

Porous Ti6Al4V scaffolds are ideal candidates as bone scaffolds since they comply with the aforementioned requirements: they are biocompatible [8,9] and their mechanical properties can be adjusted by porosity [10]. Porous structures with an interconnected pore network are of particular interest for promoting cell migration and colonization [11] as well as tissue in-growth [12,13]. Interconnected pore network enables the flow of nutrients and oxygen to cells and tissue and promotes the formation of blood capillaries [14]. When the vascular network is not sufficient, nutrient and oxygen deficiencies cause hypoxia or necrosis [15,16]. Therefore, open-porous structures with adequate pore sizes are critical for vascularization and tissue growth. However, the optimum porous structure for orthopedic scaffolds is yet to be established and there is conflicting information regarding an optimum pore size for both enhanced bone ingrowth and mechanical strength of the scaffold. Recent reviews [17–19] summarize that optimum pore size should be 300 µm or larger for bone ingrowth and vascularization. However, whilst high porosity and/or large pore size (>800 µm) promotes flow of nutrients, vascularization and tissue growth [14,15,20], highly porous structures lack the required mechanical strength and integrity and decrease the cell seeding efficiency [15,21]. At the same time, in vitro studies have shown that scaffolds with small pore sizes (<500 µm) or low porosities are prone to pore occlusions [22].

Gradient structures present a potential solution to the opposing requirements of an optimal pore size for biological response and mechanical properties. Given an optimal pore size distribution, there is potential to develop structures that exhibit both adequate mechanical strength and tissue in-growth rate. In addition, gradient structures can mimic the natural bone in terms of its structure and mechanical properties [23]. The structure of the bone changes with the amount and direction of the applied stress [24] resulting in differences in the internal structure (porosity and composition) and mechanical properties of the bone along its dimensions. For example, the elastic modulus of trabecular bone at the ends of long bones or within the interior of vertebrae is around 0.5 GPa [25]. This variation in elastic modulus depending on the location in the bone indicates the need for development of gradient structures in bone scaffolds.

The gradient and uniform porous scaffolds can be designed using traditional CAD (Computer Aided Design) and include the use of open cellular foams [26,27] and periodic uniform unit cells based on platonic solids [28–30]. Other techniques, such as implicit surface modelling [31–33] and topology optimized scaffolds [18,34], are also gaining in popularity. The fabrication of such complex structures has recently become feasible with the advances in additive manufacturing [35]. Traditional manufacturing of porous metals such as solid or liquid state processing has limited control over the shape and size of the pores achievable through adaptation of the processing parameters. These shortcomings can be overcome through additive manufacturing which builds a three-dimensional object in layer-by-layer fashion. Selective laser melting (SLM) and electron beam melting (EBM) have both been utilized to successfully fabricate porous scaffolds [36]. Both methods rely on a computer-controlled high power energy source to selectively melt a metallic powder on each layer.

A number of studies have investigated the mechanical or biological response of metallic-based gradient porous designs produced by additive manufacturing [37–45]. The majority of these studies focused on gradient structures generated by abrupt changes between layers based on change in strut diameter or unit cell volume. For instance, Li et al. [37] studied the deformation behavior of radial dual-density rhombic dodecahedron Ti6Al4V scaffolds fabricated by EBM. They achieved radial dual-density by altering the rhombic dodecahedron unit volume between two layers which resulted in discontinuity between layers. Their finite volume method simulations revealed that the inherent discontinuity between layers resulted in stress concentration and maximum stresses at the interfaces. They concluded that continuous variation between layers are ideal to minimize the stress concentration at the interface. Nune et al. [38,39] investigated the osteoblastic functions of the scaffolds designed using Li et al. [37] work, based on gradient rhombic dodecahedron created by changing the unit volume between layers. Although their work showed promising results on cellular activity when cells were seeded from large pore side (1000 µm), there was no complementary study on the mechanical properties and therefore the adverse effect of discontinuity between layers on the mechanical properties were not covered.

Another study [40] of multiple-layer gradient structures based on changing unit cell volume also showed the mismatch between two layers. They designed gradient BCC and diamond cylinders where two different unit cell volumes were used in the outer and inner parts of the cylinder. This design approach resulted in free nodes of outer layers that are not connected to inner layers, which caused a negative effect on the mechanical properties. Another approach frequently used to generate gradient structures is based on a sharp change in strut diameter between layers [41,42]. This design principle also results in a mismatch between layers negatively effecting the mechanical properties as well as tissue ingrowth and mineralization [42].

Recently, there have been a few investigations aiming to overcome the problem of a mismatch between layers by designing continuous gradient structures which consists of gradually changing strut diameters between layers. For example, Han et al. [43] reported the mechanical properties of SLM-fabricated pure titanium Schwartz diamond unit cell and demonstrated the layer-by-layer sequential failure of these gradient scaffolds. Maskery et al. [44] also showed the layer-by-layer gradual collapse of gradient scaffolds using SLM-fabricated AlSi10Mg gradient BCC structures. These two studies highlighted that the deformation and energy absorption of gradient lattices is more predictable than the uniform lattices due to the lack of diagonal shear band formation during deformation. In another study Ti6Al4V cubic and honeycomb lattice structures [45] were combined to a gradient structure with a continuous density change and it was shown that this design had a superior energy absorption properties compared to their uniform counterparts. Although the aforementioned studies on the mechanical properties of continuous gradient structures indicate promising results; the selection of unit cells, materials as well as their in vitro and in vivo response need to be better understood.

In this work, we introduce a concept of generating continuous gradient structures by changing the strut diameter linearly across cell layers which enables a smooth transition between layers. To demonstrate the benefit of this design principle, we apply it to the BCC unit cell and create gradient structures with rising or decreasing pores sizes. These gradient structures were mirrored from the central horizontal axis to obtain a symmetrical sample. Their mechanical properties were obtained by uniaxial compression tests and cell attachment and proliferation were assessed with murine pre-osteoblast cells. It will be shown that gradient scaffold can be tailored to fulfil the mechanical properties required and simultaneously improve biological response.

2. Materials and Methods

2.1. Design and Fabrication of Ti6Al4V Gradient Cellular Structures

The lattice structures were designed with Rhinoceros v5 in the RhinoPython environment. A custom-made parametric script was developed to create the lattice models with continuous gradient structures. The scaffold structure is defined by its type (BCC) (Figure 1a) and size in each of the cardinal directions, and has a changing strut diameter based upon a polynomial equation, see Equation (1). The diameter of the strut is calculated at a minimum or three locations, evenly spaced along its length based on the location of the strut, and lofted to create a smooth transition between radii.

$$Diameter = c + \sum_{n=1}^{n,(x,y,z)} A_{n_{(x,y,z)}} \left(\left| P_{(x,y,z)} \right| - P_{(x,y,z)_0} \right)^n \tag{1}$$

Here c is constant, $A_{n_{(x,y,z)}}$ is the gradient value in the current cardinal direction, $\left|P_{(x,y,z)}\right|$ is the absolute current coordinate position in the relevant axis (x, y, or z), $P_{(x,y,z)_0}$ is the gradient origin in each of the relevant axes, defined over n polynomial terms.

In the case of linear gradient in a single axis, only a single term (A_x) is required, simplifying the Equation (1) to following:

$$Diameter = c + A_x(P_x - P_{x_0}) \tag{2}$$

It should be noted that by adding y and z terms in Equation (2) will modify the nature of the gradient and allow to produce 3D gradient scaffolds.

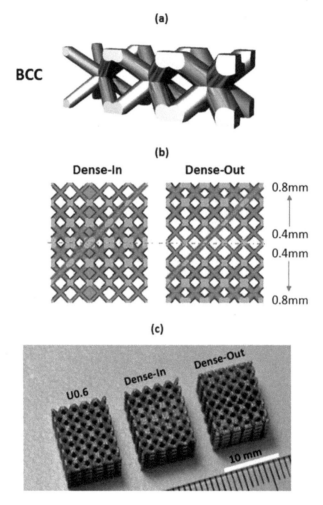

Figure 1. (a) Gradient BCC unit cell showing continuous transition at the unit cell junctions, (b) CAD view of the gradient BCC lattice scaffold; the arrows show the gradual increase in strut diameter and the highlighted areas show the gradually changing pore size and strut diameter along the scaffolds' height; Dense-In scaffold has increasing strut diameter towards the center, whereas Dense-Out follows the opposite trend; (c) SLM-fabricated specimens with dimensions of $10 \times 10 \times 12$ mm^3, from left to right: uniform BCC lattice scaffold with 0.6 mm strut diameter, Dense-In and Dense-Out scaffolds.

To demonstrate the benefit of this concept, we designed two gradient structures with dimensions of $6 \times 10 \times 10$ mm^3, mirrored in the x-axis to create lattices of $12 \times 10 \times 10$ mm^3. In this case a gradient value A_x was set to 0.07 (mm/mm) and a constant c equal to 0.4 mm. The choice of the parameters is motivated by manufacturability of the struts and leads to a minimum diameter of 0.4 mm and a maximum of 0.82 mm over the 6-mm lattice. When gradient design was mirrored from the thinner strut plane, it is named as Dense-Out and when mirrored from the thicker strut plane, it is named as Dense-In. For comparison of the mechanical and biological response of gradient structures, we also created three uniform BCC structures which utilize the same strut diameters as present in the gradient structure, namely 0.4, 0.6, 0.8 mm. These uniform BCC structures are denoted U0.4, U0.6 and U0.8, respectively.

It should be noted that the developed script is not limited to the BCC unit cell and a large library of common unit cells was programmed allowing us to generate a smooth gradient between layers and tailor their size and local and total porosity.

The uniform and gradient BCC structures were fabricated by selective laser melting (SLM) process, described in [46], using a MLab Cusing machine (Concept Laser, Lichtenfels, Germany). Ti6Al4V-ELI powder supplied by Falcon Tech Co., Ltd. (Wuxi, China) that satisfies ASTM F136 with diameter range of 15–53 μm was used for laser melting process. The Ti6Al4V samples were fabricated using a laser power of 95 W, scan speed of 600 mm/s with a hatch distance of 0.08 mm, the beam spot size of 50 (-5, $+25$) μm, with oxygen content less than 0.2% in an argon atmosphere and the layer thickness of 25 μm. The samples were built on top of a solid titanium plate and were removed by using Electrical Discharge Machining (EDM, AgieCharmilles CUT 30P, GF, Losone, Switzerland). All samples were 12 mm in height with a cross-section of 10×10 mm^2. After the samples were removed from the build plate, they were washed in an ultrasonic bath containing ethanol for 4 h to aid in removing the unmelted particles from the surface of the samples. No further surface modifications or heat treatments were applied to the scaffolds.

2.2. Morphological Analysis

Morphological properties were characterized by three methods: scanning electron microscope (SEM), digital densitometry and gas pycnometry. Pore size and strut diameter were measured using SEM (FEI Nova NanoSEM, Thermo Fisher Scientific, Hillsboro, OR, USA) images. The average of four pore sizes and strut diameters was calculated for each specimen. The volume fraction of the samples was measured by both gas pycnometry and digital densitometry. A digital densitometry (SD-200L, AlfaMirage, Osaka, Japan) adopts the Archimedean principle with liquid displacement. Three specimens of each sample were used to measure the volume fraction. In addition to densitometry measurements, the volume fraction of each sample were measured by a gas pycnometry (AccuPyc 1330, Micromeritics, Norcross, GA, USA) with helium gas in 3 repeats. The pycnometry measures the density of solid materials by employing gas displacement and Boyle's Law of gas expansion. The volume fraction values measured by gas pycnometry were compared with the values obtained from the CAD designs.

2.3. Measurement of Mechanical Properties

A minimum of five specimens of each uniform and gradient structure were mechanically tested under compression using an Instron 5982 universal testing machine with a 100 kN load cell. Following the standard for compression test for porous and cellular metals (ISO 13314:2011), a constant cross-head velocity of 0.72 mm/min was utilized corresponding to a compression strain rate of 10^{-3} s^{-1}. The measurements were recorded after a preload of 50–70 N to avoid the initial settling of the samples between plates. A series of images were captured every 1 s during compression testing to record the deformation response of the samples.

The engineering compressive stress was calculated by normalizing the applied compression load with the initial cross-section area of each sample (10×10 mm^2) and the engineering strain was calculated by the displacement of the cross-heads. The stress-strain curves of each sample were analyzed and the following mechanical properties were calculated based on the guidelines of the ISO Standard 13314:2011: first maximum compressive strength (σ_{max}) (the first local maximum in the stress-strain curve), 0.2% offset yield stress (σ_y), the elastic gradient ($E_{(\sigma 20-\sigma 50)}$) (the gradient of the elastic straight line between stresses of 20 and 50 MPa). The ISO Standard recommends determining the elastic gradient between stresses of 20 and 70 MPa; however, since 70 MPa was higher than the 0.2% yield stress point for some of the samples, 50 MPa was adopted for all samples.

2.4. Cell Viability, Proliferation and Morphology

MC3T3-E1, Subclone 4, preosteoblast cells (ATCC® CRL-2593™, Manassas, VA, USA) were cultured in α-MEM medium (Gibco) supplemented with 10% fetal bovine serum (FBS, Gibco) and 1% antibiotic/antimycotic solution (Gibco) in a humidified incubator at 37 °C, 5% CO_2. As-fabricated Ti6Al4V uniform and gradient porous structures were cleaned by successive washes with ethanol and phosphate buffer saline (PBS). Subsequently, specimens with dimensions of $10 \times 10 \times 12$ mm^3 were placed in a 24-multiwell plate with 3 repeats and seeded at a density of 1×10^5 cells per specimen. In parallel, positive control wells containing SLM-fabricated solid samples were set up. Before any evaluation, all scaffolds were transferred into a new 24-multiwell plate.

Cell viability was assessed at 4 h, 4 days and 7 days after cell culture using the MTS assay (Cell Titre 96 Aqueous One Solution Cell Proliferation Assay, Promega, Madison, WI, USA). After 2.5 h incubation with MTS reagent, 100 μL of solution from each well were transferred into a 96-multiwell plate and the optical absorbance (OD) was measured at 490 nm with a microplate reader (Thermo Scientific™ Multiskan Spectrophotometer, Vantaa, Finland). The degree of cell attachment and spreading was studied by immunofluorescence staining. Samples were fixed with 4% paraformaldehyde (PFA) for 20 min and stained with ActinRed™ 555 ReadyProbes® Reagent (Molecular Probes™, Gaithersburg, MD, USA) following the manufacturer's instructions and Hoechst 33342 (10 μg/mL) (Thermo Scientific™, Rockford, IL, USA) for 30 min. After immunostaining, the top and bottom of the specimens were examined by fluorescence microscopy (Nikon Eclipse Ti, Tokyo, Japan).

Cell morphology was characterized by scanning electron microscope (SEM) at 4 h, 4 days and 7 days of cell culture. Cells on the specimens were fixed with 2.5% glutaraldehyde in 0.1 M of sodium cacodylate buffer for 2 h and postfixed in 1% osmium tetroxide for 1.5 h at room temperature. Specimens were dehydrated with a gradient series of ethanol (50%, 70%, 80%, 90%, 95% and 100%) followed by a hexamethyldisilazane (hmds) drying procedure. The specimens were sputter coated with gold and inspected using a FEI Nova NanoSEM.

2.5. Statistical Analysis

Statistical analyses were performed using SPSS Statistics 20.0 (SPSS, Inc., Chicago, IL, USA). Data were presented as mean ± standard deviation (SD). One-way analysis of variation (ANOVA) together with Tukey–Kramer post-hoc analysis were used to identify significant differences (significance threshold: $p < 0.05$).

3. Results

3.1. Morphology of Porous Scaffolds

The key morphological parameters of the scaffolds, such as pore size and strut diameter, were measured by SEM and presented in Table 1. The measured strut diameters were larger than the original designs for all samples due to the adhesion of the semimolten powder on the surface (Figure 2b). We measured average surface roughness, R_a, using a Mitutoyo Surftest SJ400 and found it to be around 10 μm. This is in agreement with the reported values of surface roughness of SLM printed lattice structures [47]. In addition, volume fraction was quantified using the gas pycnometry and densitometry (Table 1). The volume fraction values were derived directly from the pycnometry and calculated from the densitometry, based on the following equations:

$$V = (W_a - W_w)/0.9971 \tag{3}$$

$$V_f(\%) = V/V_t \tag{4}$$

Table 1. The morphometric parameters of uniform and gradient BCC structures based on CAD designs, gas pycnometry, digital densitometry and SEM.

Scaffold Name	Design	Porosity (%)			Pore Size (mm)		Strut Diameter (mm)	
		Gas Pycnometry	Densitometry	Difference (%)	Design	SEM	Design	SEM
U0.4	82	71.87 ± 0.01	71.45 ± 0.02	13	1.51	1.14 ± 0.03	0.40	0.57 ± 0.01
U0.6	64	53.06 ± 0.01	51.11 ± 0.01	17	1.26	0.94 ± 0.05	0.60	0.77 ± 0.01
U0.8	44	33.78 ± 0.01	31.86 ± 0.01	23	1.02	0.73 ± 0.03	0.80	1.06 ± 0.02
Dense-In	62	50.73 ± 0.01	49.38 ± 0.01	18	1.33	1.04 ± 0.02	0.40	0.59 ± 0.01
					1.13	0.83 ± 0.01	0.61	0.72 ± 0.01
					0.94	0.74 ± 0.07	0.82	0.92 ± 0.03
Dense-Out	62	51.90 ± 0.02	50.01 ± 0.01	16	0.94	0.62 ± 0.02	0.82	0.91 ± 0.02
					1.13	0.82 ± 0.02	0.61	0.74 ± 0.01
					1.33	0.98 ± 0.03	0.40	0.59 ± 0.01

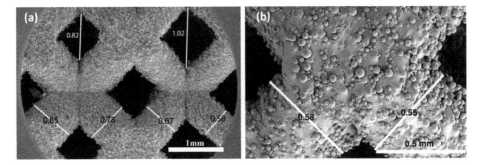

Figure 2. SEM image of (**a**) a gradient Dense-Out structure, demonstrating the change in strut diameter and pore size along the scaffold, (**b**) higher magnification view of the struts showing attached semi-molten powders.

Here, V is the volume of the scaffolds, W_a is the weight measured in air, W_w is the weight measured in water, 0.9971 is the density (g cm^{-3}) of distilled water at 25 °C and 1 atm, V_f is the volume fraction in percent, V_t is the total volume obtained from the outer dimension of the scaffolds, which is 1.2 cm^3 for all. The porosity of the scaffolds is defined as 100-Volume fraction (%).

The measured values of volume fraction were smaller than the designed volume fraction values for all samples. The difference between the original designs and pyconometry results was around 13–23%. These deviations are as expected since the as-fabricated strut diameters were larger than the original designs, resulting in smaller pore sizes than the intended geometry. This difference is characteristic to SLM process and is caused by effects such as staircase stepping due to layered manufacturing and melt pool variation due to residual stresses [48].

Strut diameter change along the gradient BCC scaffolds was noticeable in SEM images, (Figure 2a), confirming that the desired graded porosity was successfully achieved by the SLM process. The pore size of gradient Dense-In scaffold was varied from 1.14 mm to 0.74 mm, whereas it was between

0.62 mm to 0.98 mm for gradient Dense-Out scaffold. Measured porosity of the gradient Dense-In and Dense-Out scaffolds were almost identical due to symmetric design along the horizontal center plane.

3.2. Mechanical Properties of Porous Scaffolds

The compressive nominal stress–strain plots of uniform and gradient BCC structures are presented in Figure 3. The stress–strain curves exhibit characteristic stages of deformation for cellular solids [10,49], including linear elastic region, followed by plateau region with fluctuating stresses. The uniform structures showed similar behavior under compression; however, they reached different levels of maximum stress and possess different elastic moduli. The stress–strain curves for gradient structures also showed initially similar behavior to the uniform scaffolds. After the onset of plasticity an abrupt structural collapse was observed in uniform scaffolds, but not in the gradient scaffolds. Furthermore, the fluctuating degree of plateau region was more distinguished for the uniform scaffolds than the gradient scaffolds.

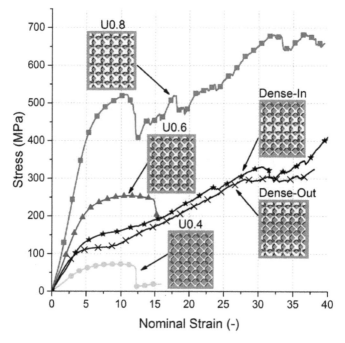

Figure 3. Nominal stress–strain curves for uniform and gradient structures.

The elastic gradient between stresses of 20 MPa and 50 MPa ($E_{(\sigma20-\sigma50)}$), the 0.2% offset yield stress (σ_y) and the first maximum compressive strength (σ_{max}) of the scaffolds are summarized (Table 2). Elastic modulus, yield stress and compressive strength increases with extension in strut diameter of the uniform structures or decrease in porosity. Aligned with the expectations of composite rule of mixtures [50], the values of elastic modulus, compressive strength and yield stress of gradient structures lie between those values of the uniform structures. No significant difference between elastic modulus of gradient structures and Uniform 0.6 sample was observed (Supplementary document Figure S1).

Table 2. The summary of the mechanical properties of uniform and gradient BCC structures measured by compression tests. (Mean ± SD).

Scaffold Name	$E_{\sigma 20-\sigma 50}$ (GPa)	σ_y (MPa)	σ_{max} (MPa)
U0.4	1.6 ± 0.2	53 ± 4	74 ± 2
U0.6	4.6 ± 0.4	192 ± 14	256 ± 4
U0.8	9.0 ± 0.6	392 ± 14	532 ± 11
Dense-In	3.9 ± 0.8	114 ± 8	150 ± 17
Dense-Out	3.5 ± 0.5	86 ± 11	128 ± 8

Relative modulus (E/E_0) against the measured volume fraction (%) is plotted in Figure 4. Elastic modulus was normalized relative to the values of solid Ti6Al4V (110 GPa). The observed average trend shows a positive power relation with the volume fraction and this trend corresponds to theoretically expected behavior of bending-dominated structures [10,51]. Results of a similar study from the literature were used to support our data [52]. The gradient structures were not considered for the power law curve since their volume fraction is similar to U0.6 specimen.

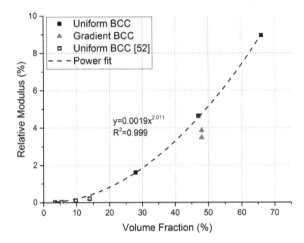

Figure 4. Relative elastic modulus vs volume fraction (%) of uniform and gradient BCC structures. Power law curve and equation was fitted on the uniform BCC structure data and demonstrates bending-dominated behavior.

Images of the initial stage and the progressive failure of uniform and gradient structures recorded during the compression tests (Figure 5) show that the major failure bands were formed at a 45° angle from the loading direction for all uniform BCC structures. For the gradient structures, the fracture initiated from the thinnest struts, that is at the top and bottom plane for Dense-In and in the middle for Dense-Out. This diagonal shear collapse of uniform structures is typical behavior of BCC structures [52–54] and other structures with different cell geometries [55,56] owing to strut bending at lattice joints [51].

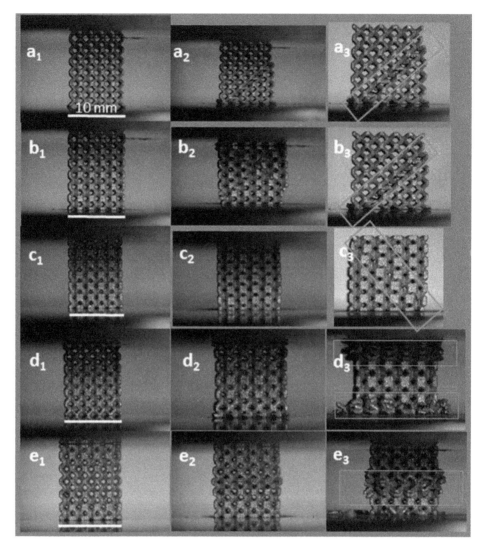

Figure 5. Failure modes of (**a**) U0.4, (**b**) U0.6, (**c**) U0.8, (**d**) Dense-In, (**e**) Dense-Out structures. Left images (subscript 1) represent the initial state and middle (subscript 2) and last right images (subscript 3) present the progressive failure. Highlights represent the observed regions of deformation and failure. (Scale bars = 10 mm).

3.3. Cellular Response to Porous Scaffolds

In order to determine the ability of the scaffolds to interact with cells, the adhesion of MC3T3-E1 preosteoblast cells on the gradient, uniform and solid scaffolds were determined by MTS assay after 4 h of incubation and showed no significant difference in cell seeding between the scaffolds (Figure 6).

In order to determine the extent of cell proliferation on the scaffolds, an MTS assay was performed after 4 and 7 days of culture. The uniform scaffolds showed a trend of decreasing cell number with increasing strut diameter at both days 4 and 7 (Figure 7). For the U0.4 and U0.6 scaffolds the number of cells approximately doubled across this time period whilst there was only a 70% increase on the

U0.8 scaffold (Figure 8), further extending the difference in cell number between the samples. For the gradient scaffolds, there were no significant differences in cell number on day 4. However, from day 4 to day 7, there was almost a 400% increase in cell numbers on the Dense-In scaffold but only 20% increase in cell numbers on the Dense-Out scaffold, resulting in significantly fewer cells on the Dense-Out scaffold as compared to the Dense-In scaffold on day 7. Although solid control sample showed the highest cell number at day 7, the percentage increase from day 4 to day 7 was largest for Dense-In scaffold. The final cell number on the Dense-Out scaffold was comparable to that of the U0.8, with both scaffolds having similar diameter of the outermost struts of the design. These results suggest that the scaffolds having thinner struts or larger pores on their outside surface (such as U0.4 and Dense-In) were more favorable for cell proliferation than the scaffolds having thicker struts or smaller pores on their outer surface. Given that the surface area of Dense-In and Dense-Out is identical as a result of their symmetrical design, it can be said that the cell viability was independent of surface area in this study.

Further to cell proliferation, cell distribution on the uniform and gradient scaffolds was studied by staining and imaging the cell nuclei and actin cytoskeleton. After 4 h of incubation, all of the scaffolds had similar cell distribution on their top surface (i.e., the surface onto which the cells were seeded) (Figure 9). The lack of cells at the bottom of the scaffolds suggests that most of the initial attachment was on the top surface. After 4 days, substantially more cells were observed both on the top and bottom surfaces of the U0.4 and Dense-In scaffolds, whereas there were no noticeable differences on the other scaffolds between 4 h and day 4 time points (Figure 10). At day 7, all the scaffolds had high density of cells on their top surface; whilst, the bottom surface of U0.6, U0.8 and Dense-Out scaffolds had almost no cells. In contrast, the bottom surface of U0.4 and Dense-In scaffolds had visibly higher cell densities.

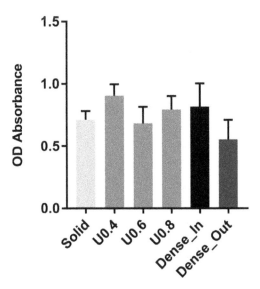

Figure 6. Adhesion of cells to uniform and gradient porous structures and to solid control sample as measured by MTS assay after 4 h. The optical absorbance (OD) was measured at 490 nm. Data are presented as mean $\pm$ SD (n = 3). No statistically significant differences were observed between scaffolds.

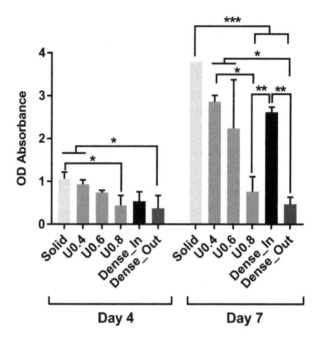

Figure 7. Cell proliferation measured by MTS assay after culturing 4 and 7 days on the uniform and gradient porous structures. The optical absorbance (OD) was measured at 490 nm. Data were presented as mean ± SD ($n = 3$). (* $p < 0.05$, ** $p < 0.01$, *** $p < 0.001$ when compared using ANOVA Tukey–Kramer post-hoc test).

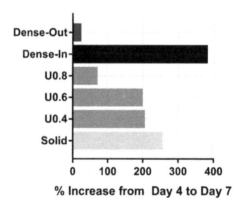

Figure 8. Percentage increase in od absorbance of the scaffolds from day 4 to day 7.

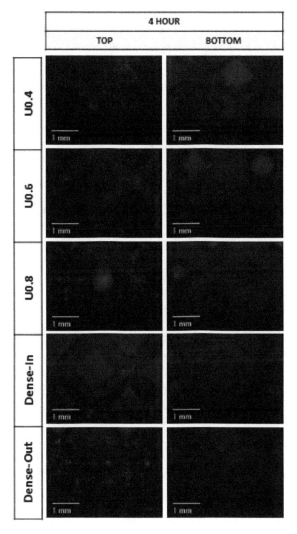

Figure 9. Fluorescence micrographs representing merged Hoechst stained nucleus (blue) and actin cytoskeleton (red) of MC3T3-E1 preosteoblast cells on the uniform and gradient BCC structures after culturing for 4 h. Top represents the side where cells were seeded onto the samples.

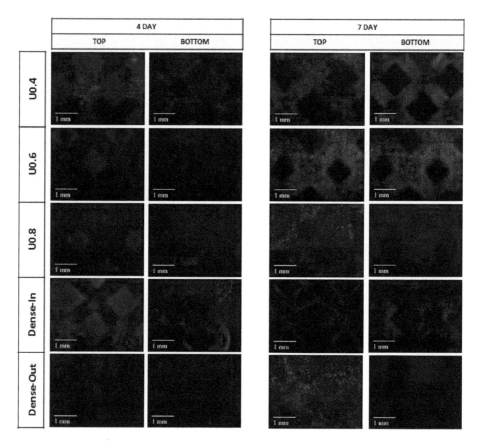

Figure 10. Fluorescence micrographs representing merged Hoechst stained nucleus (blue) and actin cytoskeleton (red) of MC3T3-E1 preosteoblast cells on the uniform and gradient BCC structures after culturing for 4 days and 7 days. Top represents the side where cells were seeded onto the samples.

Despite the difference in cell proliferation and migration on the different scaffolds, cell morphology was similar for the scaffolds when the images were taken from the top surface (Figure 9). SEM images taken from the middle and bottom part of the scaffolds supported the findings of fluorescent images showing differing cell penetration depth profiles for the varying scaffold structures (Supplementary document Figures S2–S6). The number of cells decreased in the middle and bottom parts of the U0.8 and Dense-Out scaffolds, as compared to U0.4 and Dense-In scaffolds. Cells were noticed to form a sheet-like elongated matrix (dashed line) (Figure 9). Moreover, the morphology of cells presented a high density of filopodia-like projections (red arrows) extending from the leading edges of cells and interacting with the substrate. The interaction between cells and substrate observed by SEM (Figure 11) shows that cells attach both on and between the unmelted powders, indicating that SLM process are beneficial to cell attachment and colonization.

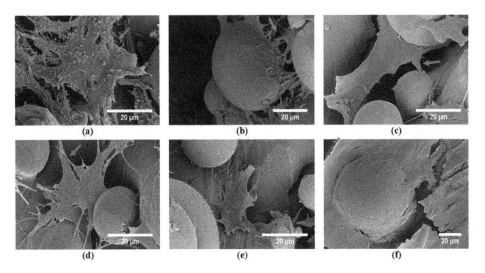

Figure 11. SEM images of the MC3T3 preosteoblast cells after culturing for 7 days on the top surfaces of (**a**) U0.4, (**b**) U0.6, (**c**) U0.8, (**d**) Dense-In, (**e**) Dense-Out, (**f**) solid scaffolds.

4. Discussion

In this work, the effect of gradient porous structures on both biological response and mechanical behavior is investigated. The generated gradient structures, denoted as Dense-In and Dense-Out, utilize gradual change in diameter and therefore minimize the stress concentration at the lattice junctions. The designed pore sizes changed from 940 μm to 1330 μm for the gradient scaffolds. The deviation of pore size and strut diameter between CAD design and fabricated scaffolds were within an expected range [57] and was attributed to surface irregularities [58]. The deviation for each scaffold was similar, demonstrating the consistency of the SLM fabrication process. Porosities of the uniform scaffolds varied from 32% to 72%, and both gradient structures had a porosity of 50%. SEM images revealed that all scaffolds had unmelted powder attached to the surface of the struts due to layered manufacturing and melt pool variation during the SLM process.

The stiffness of the tested scaffolds varied in the range of 1.6 to 9.0 GPa, which aligns with the stiffness range of the trabecular (0.4 GPa [59,60]) and cortical bones (3–20 GPa [5,24]). The yield stress values of the scaffolds were in the range of 53 to 392 MPa, which lies in the range of cortical bones (33–193 MPa [24,61]), but it is not suitable for a replacement of trabecular bone, 2–17 MPa [60]. Considering the scaffolds presented in this study aim to be used as load-bearing implants for replacement of cortical bones, the yield stress and elastic modulus values satisfy the mechanical property requirements.

Table 2 shows that the elastic modulus, yield stress and maximum compressive strength values increase as the strut diameter increases or porosity decreases. The mechanical properties of gradient structures lie in the range of representative values of uniform structures and can be predicted based on an assumption that gradient structures are composites of uniform layers of the same diameter struts. Based on this assumption, the elastic modulus of gradient structure in uniaxial compression can be calculated through the general rule of mixtures [42,50]:

$$\frac{1}{E_{Gradient}} = \frac{1}{3E_{U0.4}} + \frac{1}{3E_{U0.6}} + \frac{1}{3E_{U0.8}} \tag{5}$$

Using Equation (1), elastic modulus of gradient structure was calculated to be 3.2 GPa, which is comparable to the measured values of 3.9 and 3.5 GPa for Dense-In and Dense-Out scaffolds, respectively.

The deformation response of uniform BCC structures follows the bending-dominated behavior with diagonal shear collapse. Interestingly, the failure mechanism of gradient structures was different due to sequential layer collapse and various deformation stages occurring simultaneously. Thinner struts reached the densification stage (when two opposite cell walls come together as the pore size decreases) while the thicker struts were still in the plateau region during the compression test. The predominant fracture band of gradient structures was initiated at the thinnest struts due to high stress concentrations on the thin strut junctions. This failure mechanism has also been noted in other studies [44,45].

Further to mechanical behavior studies, we analyzed the in vitro response of the scaffolds with preosteoblast cells. The degree of cell attachment was similar for all scaffolds, but cell proliferation and colonization were significantly different. Scaffolds with a thin strut diameter on the periphery (U0.4 and Dense-In) allowed cells to populate throughout the scaffold whereas those with a thicker outer strut (U0.6, U0.8 and Dense-Out) did not allow cells to migrate to the bottom surface, suggesting that cells were entrapped at the smaller pore size region (top surface). Consequently, the proliferation rate of cells on these scaffolds was markedly less. Although this immobilization behavior of the preosteoblast cells when seeded from small pore size region was observed in the previous studies of Nune et al. [38,39], the underlying reason for this behavior is still unknown. The smallest pore size in our scaffolds were 940 μm which is larger than the suggested pore size (100–300 μm) for cell colonization and migration [62,63]. In addition, the smallest pore size is much larger than an average size of a MC3T3-E1 pre-osteoblast cells (20–40 μm, Figure 9). It is not yet clear why thicker struts on the scaffold periphery inhibit cell activity whilst the thick struts on the interior of the Dense-In scaffold did not deter colonization of cells through the whole structure.

Our results suggest that the surface area does not affect the cell attachment and proliferation. The large surface area of the U0.8 scaffold, due to its large strut diameters, was expected to promote cell attachment and growth; however, it showed the lowest cell number at day 7. Similar behavior was observed for the gradient structures, which possessed equal surface area but showed a large difference in cell number. It is therefore likely that, parameters other than the surface area of the scaffold affected the cell colonization. Identification of the specific factors would require further clarification but could include the flow conditions and cellular aggregation [39]. In addition, it would be interesting to assess whether vascularization happens more quickly or easily with larger pores on the periphery than the smaller pores on the periphery.

In recent years, additively manufactured gradient structures for tissue engineering have been studied [37–39,41,44,64,65]; however these studies either focused on the mechanical properties or biological response independently. In this work, the biological and mechanical responses were assessed simultaneously allowing us to study the overall impact of the designed geometry of the scaffolds. Our results suggest that when designing a porous gradient structure, both biological and mechanical requirements must be considered concurrently, since their requirements are opposing. The compression test results demonstrate the benefit of utilizing smaller pore size in increasing the stiffness and strength of the porous scaffolds; whereas, the cell proliferation data suggests that scaffolds with larger pore size in their outer surface favors cell proliferation. Therefore, it can be concluded that gradient scaffolds provide a possible solution for overcoming the conflicting requirements of bone tissue implants. Gradient structures with decreasing pore size towards their center can provide the required strength and stiffness, while simultaneously promoting cell colonization throughout the whole scaffold. The Dense-In scaffold fabricated in this study has an elastic modulus of 3.9 GPa which is in the range for those of cortical bone [5]. Furthermore, this scaffold has a varying pore size ranging from 1330 μm on the outside to 940 μm at the core. These values have been previously reported to be favorable for cell colonization as well as bone ingrowth and vascularization [66–68].

In summary, an ideal scaffold for bone regeneration should facilitate cell attachment, infiltration and matrix deposition to guide bone formation [69] as well as providing initial mechanical support to the surrounding bone [70]. Porous titanium scaffolds can meet the mechanical strength and bone formation requirements without osseoinductive biomolecules [68]; however, the pore size of the scaffolds needs to be high for bone-ingrowth whereas, as the porosity increases, the mechanical strength and integrity of the structure decreases [71]. Gradient structures represent an ideal candidate to overcome these opposing requirements of high porosity and mechanical strength.

Our study demonstrated the benefit of the gradient scaffold with larger pores in its outer surface in terms of gaining optimum mechanical strength and promoting cell attachment and colonization. In addition, this framework demonstrates that mechanical properties can be tailored through gradient structure design and simultaneously improve the biological response. This approach therefore holds significant promise in the development of orthopedic implants, where the location of the implant and the corresponding loading condition can dictate the implant topology.

5. Conclusions

This work combined and assessed the in vitro behavior and mechanical response of gradient and uniform porous scaffolds for bone tissue engineering. For this purpose, five different BCC structures were fabricated using selective laser melting technology. Static mechanical properties of the gradient structures followed the rule of mixtures and the obtained values are in the range of those corresponding values for uniform structures. The mechanical properties of all studied scaffolds are comparable to the reported mechanical properties of the cortical bone. Quantitative analysis of cell viability showed higher cell colonization and proliferation rates for scaffolds with large pores (1000–1100 μm) in their outer surface after 4 and 7 days of culturing. However, when comparing the mechanical properties of structures with this comparable biological activity, the uniform U0.4 scaffold showed less than half of the respective mechanical properties for the Dense-In scaffold. The combined results of compression tests and in vitro biological analyses indicate that the Dense-In scaffold is an ideal porous structure to balance mechanical and biological performances to meet the requirements of load-bearing implants. Based on the results presented in this work, optimal gradient structures should possess small pores in their core in order to increase their mechanical integrity and strength while large pores should be utilized in their outer surface to avoid pore occlusion. We suggest that this approach could be widely used in the design of orthopedic implants to maximize both the mechanical and biological properties of the implant.

Supplementary Materials: The following are available online at http://www.mdpi.com/2075-4701/8/4/200/s1, Figure S1: Mechanical properties of the scaffolds; Figures S2–S6: Observation of cell morphology and distribution along the scaffolds by SEM.

Acknowledgments: Ezgi Onal would like to acknowledge The Clive and Vera Ramaciotti Centre for Structural Cryo-Electron Microscopy. This project is funded by the ARC Research Hub for Transforming Australia's Manufacturing Industry through High Value Additive Manufacturing (IH130100008) and Jessica E. Frith is supported by an ARC DECRA (DE130100986).

Author Contributions: E.O. and A.M. conceived and designed the experiments; E.O. performed the experiments; E.O., J.E.F., X.W. and A.M. analyzed the data; M.J. contributed generating Python scripts to design structures; all authors contributed to the writing of the paper.

Conflicts of Interest: The authors declare no conflicts of interest.

References

1. Hutmacher, D.W. Scaffolds in tissue engineering bone and cartilage. *Biomaterials* **2000**, *21*, 2529–2543. [CrossRef]
2. Hutmacher, D.W.; Schantz, J.T.; Lam, C.X.F.; Tan, K.C.; Lim, T.C. State of the art and future directions of scaffold-based bone engineering from a biomaterials perspective. *J. Tissue Eng. Regener. Med.* **2007**, *1*, 245–260. [CrossRef] [PubMed]

3. Niinomi, M.; Nakai, M.; Hieda, J. Development of new metallic alloys for biomedical applications. *Acta Biomater.* **2012**, *8*, 3888–3903. [CrossRef] [PubMed]

4. Andani, M.T.; Shayesteh Moghaddam, N.; Haberland, C.; Dean, D.; Miller, M.J.; Elahinia, M. Metals for bone implants. Part 1. Powder metallurgy and implant rendering. *Acta Biomater.* **2014**, *10*, 4058–4070. [CrossRef] [PubMed]

5. Bayraktar, H.H.; Morgan, E.F.; Niebur, G.L.; Morris, G.E.; Wong, E.K.; Keaveny, T.M. Comparison of the elastic and yield properties of human femoral trabecular and cortical bone tissue. *J. Biomech.* **2004**, *37*, 27–35. [CrossRef]

6. Moghaddam, N.S.; Andani, M.T.; Amerinatanzi, A.; Haberland, C.; Huff, S.; Miller, M.; Elahinia, M.; Dean, D. Metals for bone implants: Safety, design, and efficacy. *Biomanuf. Rev.* **2016**, *1*, 1. [CrossRef]

7. Al-Tamimi, A.A.; Fernandes, P.R.A.; Peach, C.; Cooper, G.; Diver, C.; Bartolo, P.J. Metallic bone fixation implants: A novel design approach for reducing the stress shielding phenomenon. *Virtual Phys. Prototyp.* **2017**, *12*, 141–151. [CrossRef]

8. Long, M.; Rack, H.J. Titanium alloys in total joint replacement—A materials science perspective. *Biomaterials* **1998**, *19*, 1621–1639. [CrossRef]

9. Abdel-Hady Gepreel, M.; Niinomi, M. Biocompatibility of Ti-alloys for long-term implantation. *J. Mech. Behav. Biomed. Mater.* **2013**, *20*, 407–415. [CrossRef] [PubMed]

10. Ashby, M.F. The properties of foams and lattices. *Philos. Trans. R. Soc. A Math. Phys. Eng. Sci.* **2006**, *364*, 15–30. [CrossRef] [PubMed]

11. Alvarez, K.; Nakajima, H. Metallic scaffolds for bone regeneration. *Materials* **2009**, *2*, 790. [CrossRef]

12. Heinl, P.; Müller, L.; Körner, C.; Singer, R.F.; Müller, F.A. Cellular Ti–6Al–4V structures with interconnected macro porosity for bone implants fabricated by selective electron beam melting. *Acta Biomater.* **2008**, *4*, 1536–1544. [CrossRef] [PubMed]

13. Marin, E.; Fusi, S.; Pressacco, M.; Paussa, L.; Fedrizzi, L. Characterization of cellular solids in Ti6Al4V for orthopedic implant applications: Trabecular titanium. *J. Mech. Behav. Biomed. Mater.* **2010**, *3*, 373–381. [CrossRef] [PubMed]

14. Rouwkema, J.; Rivron, N.C.; van Blitterswijk, C.A. Vascularization in tissue engineering. *Trends Biotechnol.* **2008**, *26*, 434–441. [CrossRef] [PubMed]

15. Kumar, A.; Nune, K.C.; Murr, L.E.; Misra, R.D.K. Biocompatibility and mechanical behaviour of three-dimensional scaffolds for biomedical devices: Process–structure–property paradigm. *Int. Mater. Rev.* **2016**, *61*, 20–45. [CrossRef]

16. Bramfeld, H.; Sabra, G.; Centis, V.; Vermette, P. Scaffold vascularization: A challenge for three-dimensional tissue engineering. *Curr. Med. Chem.* **2010**, *17*, 3944–3967. [CrossRef]

17. Perez, R.A.; Mestres, G. Role of pore size and morphology in musculo-skeletal tissue regeneration. *Mater. Sci. Eng. C* **2016**, *61*, 922–939. [CrossRef] [PubMed]

18. Wang, X.; Xu, S.; Zhou, S.; Xu, W.; Leary, M.; Choong, P.; Qian, M.; Brandt, M.; Xie, Y.M. Topological design and additive manufacturing of porous metals for bone scaffolds and orthopedic implants: A review. *Biomaterials* **2016**, *83*, 127–141. [CrossRef] [PubMed]

19. Sing, S.L.; An, J.; Yeong, W.Y.; Wiria, F.E. Laser and electron-beam powder-bed additive manufacturing of metallic implants: A review on processes, materials and designs. *J. Orthop. Res.* **2016**, *34*, 369–385. [CrossRef] [PubMed]

20. Li, J.P.; Habibovic, P.; van den Doel, M.; Wilson, C.E.; de Wijn, J.R.; van Blitterswijk, C.A.; de Groot, K. Bone ingrowth in porous titanium implants produced by 3D fiber deposition. *Biomaterials* **2007**, *28*, 2810–2820. [CrossRef] [PubMed]

21. Warnke, P.H.; Douglas, T.; Wollny, P.; Sherry, E.; Steiner, M.; Galonska, S.; Becker, S.T.; Springer, I.N.; Wiltfang, J.; Sivananthan, S. Rapid prototyping: Porous titanium alloy scaffolds produced by selective laser melting for bone tissue engineering. *Tissue Eng. Part C Methods* **2008**, *15*, 115–124. [CrossRef] [PubMed]

22. Van Bael, S.; Chai, Y.C.; Truscello, S.; Moesen, M.; Kerckhofs, G.; Van Oosterwyck, H.; Kruth, J.P.; Schrooten, J. The effect of pore geometry on the in vitro biological behavior of human periosteum-derived cells seeded on selective laser-melted Ti6Al4V bone scaffolds. *Acta Biomater.* **2012**, *8*, 2824–2834. [CrossRef] [PubMed]

23. Leong, K.F.; Chua, C.K.; Sudarmadji, N.; Yeong, W.Y. Engineering functionally graded tissue engineering scaffolds. *J. Mech. Behav. Biomed. Mater.* **2008**, *1*, 140–152. [CrossRef] [PubMed]

24. Karageorgiou, V.; Kaplan, D. Porosity of 3D biomaterial scaffolds and osteogenesis. *Biomaterials* **2005**, *26*, 5474–5491. [CrossRef] [PubMed]

25. Li, G.; Wang, L.; Pan, W.; Yang, F.; Jiang, W.; Wu, X.; Kong, X.; Dai, K.; Hao, Y. In vitro and in vivo study of additive manufactured porous Ti6Al4V scaffolds for repairing bone defects. *Sci. Rep.* **2016**, *6*, 34072. [CrossRef] [PubMed]

26. Murr, L.E.; Gaytan, S.M.; Medina, F.; Martinez, E.; Martinez, J.L.; Hernandez, D.H.; Machado, B.I.; Ramirez, D.A.; Wicker, R.B. Characterization of Ti–6Al–4V open cellular foams fabricated by additive manufacturing using electron beam melting. *Mater. Sci. Eng. A* **2010**, *527*, 1861–1868. [CrossRef]

27. Murr, L.E.; Gaytan, S.M.; Medina, F.; Lopez, H.; Martinez, E.; Machado, B.I.; Hernandez, D.H.; Martinez, L.; Lopez, M.I.; Wicker, R.B.; et al. Next-generation biomedical implants using additive manufacturing of complex, cellular and functional mesh arrays. *Philos. Trans. R. Soc. Lond. A Math. Phys. Eng. Sci.* **2010**, *368*, 1999–2032. [CrossRef] [PubMed]

28. Arabnejad, S.; Burnett Johnston, R.; Pura, J.A.; Singh, B.; Tanzer, M.; Pasini, D. High-strength porous biomaterials for bone replacement: A strategy to assess the interplay between cell morphology, mechanical properties, bone ingrowth and manufacturing constraints. *Acta Biomater.* **2016**, *30*, 345–356. [CrossRef] [PubMed]

29. Wettergreen, M.A.; Bucklen, B.S.; Starly, B.; Yuksel, E.; Sun, W.; Liebschner, M.A.K. Creation of a unit block library of architectures for use in assembled scaffold engineering. *Comput.-Aided Des.* **2005**, *37*, 1141–1149. [CrossRef]

30. Parthasarathy, J.; Starly, B.; Raman, S. A design for the additive manufacture of functionally graded porous structures with tailored mechanical properties for biomedical applications. *J. Manuf. Process.* **2011**, *13*, 160–170. [CrossRef]

31. Bobbert, F.S.L.; Lietaert, K.; Eftekhari, A.A.; Pouran, B.; Ahmadi, S.M.; Weinans, H.; Zadpoor, A.A. Additively manufactured metallic porous biomaterials based on minimal surfaces: A unique combination of topological, mechanical, and mass transport properties. *Acta Biomater.* **2017**, *53*, 572–584. [CrossRef] [PubMed]

32. Giannitelli, S.M.; Accoto, D.; Trombetta, M.; Rainer, A. Current trends in the design of scaffolds for computer-aided tissue engineering. *Acta Biomater.* **2014**, *10*, 580–594. [CrossRef] [PubMed]

33. Kapfer, S.C.; Hyde, S.T.; Mecke, K.; Arns, C.H.; Schröder-Turk, G.E. Minimal surface scaffold designs for tissue engineering. *Biomaterials* **2011**, *32*, 6875–6882. [CrossRef] [PubMed]

34. Zhang, X.-Y.; Fang, G.; Zhou, J. Additively manufactured scaffolds for bone tissue engineering and the prediction of their mechanical behavior: A review. *Materials* **2017**, *10*, 50. [CrossRef] [PubMed]

35. Horn, T.J.; Harrysson, O.L.A. Overview of current additive manufacturing technologies and selected applications. *Sci. Prog.* **2012**, *95*, 255–282. [CrossRef] [PubMed]

36. Sidambe, A. Biocompatibility of advanced manufactured titanium implants—A review. *Materials* **2014**, *7*, 8168–8188. [CrossRef] [PubMed]

37. Li, S.; Zhao, S.; Hou, W.; Teng, C.; Hao, Y.; Li, Y.; Yang, R.; Misra, R.D.K. Functionally graded Ti-6Al-4V meshes with high strength and energy absorption. *Adv. Eng. Mater.* **2016**, *18*, 34–38. [CrossRef]

38. Nune, K.; Kumar, A.; Misra, R.; Li, S.; Hao, Y.; Yang, R. Osteoblast functions in functionally graded Ti-6Al-4V mesh structures. *J. Biomater. Appl.* **2016**, *30*, 1182–1204. [CrossRef] [PubMed]

39. Nune, K.C.; Kumar, A.; Misra, R.D.K.; Li, S.J.; Hao, Y.L.; Yang, R. Functional response of osteoblasts in functionally gradient titanium alloy mesh arrays processed by 3D additive manufacturing. *Colloids Surf. B Biointerfaces* **2017**, *150*, 78–88. [CrossRef] [PubMed]

40. Surmeneva, M.A.; Surmenev, R.A.; Chudinova, E.A.; Koptioug, A.; Tkachev, M.S.; Gorodzha, S.N.; Rännar, L.-E. Fabrication of multiple-layered gradient cellular metal scaffold via electron beam melting for segmental bone reconstruction. *Mater. Des.* **2017**, *133*, 195–204. [CrossRef]

41. Limmahakhun, S.; Oloyede, A.; Sitthiseripratip, K.; Xiao, Y.; Yan, C. Stiffness and strength tailoring of cobalt chromium graded cellular structures for stress-shielding reduction. *Mater. Des.* **2017**, *114*, 633–641. [CrossRef]

42. Van Grunsven, W.; Hernandez-Nava, E.; Reilly, G.; Goodall, R. Fabrication and mechanical characterisation of titanium lattices with graded porosity. *Metals* **2014**, *4*, 401–409. [CrossRef]

43. Han, C.; Li, Y.; Wang, Q.; Wen, S.; Wei, Q.; Yan, C.; Hao, L.; Liu, J.; Shi, Y. Continuous functionally graded porous titanium scaffolds manufactured by selective laser melting for bone implants. *J. Mech. Behav. Biomed. Mater.* **2018**, *80*, 119–127. [CrossRef] [PubMed]

44. Maskery, I.; Aboulkhair, N.T.; Aremu, A.O.; Tuck, C.J.; Ashcroft, I.A.; Wildman, R.D.; Hague, R.J.M. A mechanical property evaluation of graded density Al-Si10-Mg lattice structures manufactured by selective laser melting. *Mater. Sci. Eng. A* **2016**, *670*, 264–274. [CrossRef]

45. Choy, S.Y.; Sun, C.-N.; Leong, K.F.; Wei, J. Compressive properties of functionally graded lattice structures manufactured by selective laser melting. *Mater. Des.* **2017**, *131*, 112–120. [CrossRef]

46. Yap, C.Y.; Chua, C.K.; Dong, Z.L.; Liu, Z.H.; Zhang, D.Q.; Loh, L.E.; Sing, S.L. Review of selective laser melting: Materials and applications. *Appl. Phys. Rev.* **2015**, *2*, 041101. [CrossRef]

47. Pyka, G.; Kerckhofs, G.; Papantoniou, I.; Speirs, M.; Schrooten, J.; Wevers, M. Surface Roughness and Morphology Customization of Additive Manufactured Open Porous Ti6Al4V Structures. *Materials* **2013**, *6*, 4737–4757. [CrossRef] [PubMed]

48. Wang, D.; Yang, Y.; Liu, R.; Xiao, D.; Sun, J. Study on the designing rules and processability of porous structure based on selective laser melting (SLM). *J. Mater. Process. Technol.* **2013**, *213*, 1734–1742. [CrossRef]

49. Rashed, M.G.; Ashraf, M.; Mines, R.A.W.; Hazell, P.J. Metallic microlattice materials: A current state of the art on manufacturing, mechanical properties and applications. *Mater. Des.* **2016**, *95*, 518–533. [CrossRef]

50. Nemat-Nasser, S.; Hori, M. *Micromechanics: Overall Properties of Heterogeneous Materials*, 2nd ed.; Elsevier: Amsterdam, The Netherlands, 1998.

51. Mazur, M.; Leary, M.; Sun, S.; Vcelka, M.; Shidid, D.; Brandt, M. Deformation and failure behaviour of Ti-6Al-4V lattice structures manufactured by selective laser melting (SLM). *Int. J. Adv. Manuf. Technol.* **2016**, *84*, 1391–1411. [CrossRef]

52. Smith, M..; Guan, Z.; Cantwell, W.J. Finite element modelling of the compressive response of lattice structures manufactured using the selective laser melting technique. *Int. J. Mech. Sci.* **2013**, *67*, 28–41. [CrossRef]

53. Gorny, B.; Niendorf, T.; Lackmann, J.; Thoene, M.; Troester, T.; Maier, H.J. In situ characterization of the deformation and failure behavior of non-stochastic porous structures processed by selective laser melting. *Mater. Sci. Eng. A* **2011**, *528*, 7962–7967. [CrossRef]

54. Cansizoglu, O.; Harrysson, O.; Cormier, D.; West, H.; Mahale, T. Properties of Ti-6Al-4V non-stochastic lattice structures fabricated via electron beam melting. *Mater. Sci. Eng. A* **2008**, *492*, 468–474. [CrossRef]

55. Zhao, S.; Li, S.J.; Hou, W.T.; Hao, Y.L.; Yang, R.; Misra, R.D.K. The influence of cell morphology on the compressive fatigue behavior of Ti-6Al-4V meshes fabricated by electron beam melting. *J. Mech. Behav. Biomed. Mater.* **2016**, *59*, 251–264. [CrossRef] [PubMed]

56. Li, S.J.; Xu, Q.S.; Wang, Z.; Hou, W.T.; Hao, Y.L.; Yang, R.; Murr, L.E. Influence of cell shape on mechanical properties of Ti-6Al-4V meshes fabricated by electron beam melting method. *Acta Biomater.* **2014**, *10*, 4537–4547. [CrossRef] [PubMed]

57. Van Bael, S.; Kerckhofs, G.; Moesen, M.; Pyka, G.; Schrooten, J.; Kruth, J.P. Micro-CT-based improvement of geometrical and mechanical controllability of selective laser melted Ti6Al4V porous structures. *Mater. Sci. Eng. A* **2011**, *528*, 7423–7431. [CrossRef]

58. Parthasarathy, J.; Starly, B.; Raman, S.; Christensen, A. Mechanical evaluation of porous titanium (Ti6Al4V) structures with electron beam melting (EBM). *J. Mech. Behav. Biomed. Mater.* **2010**, *3*, 249–259. [CrossRef] [PubMed]

59. Linde, F.; Hvid, I. The effect of constraint on the mechanical behaviour of trabecular bone specimens. *J. Biomech.* **1989**, *22*, 485–490. [CrossRef]

60. Morgan, E.F.; Keaveny, T.M. Dependence of yield strain of human trabecular bone on anatomic site. *J. Biomech.* **2001**, *34*, 569–577. [CrossRef]

61. Cullinane, D.M.; Einhorn, T.A. Biomechanics of bone. In *Principles of Bone Biology*, 2nd ed.; Raisz, L.G., Rodan, G.A., Eds.; Academic Press: San Diego, CA, USA, 2002.

62. Murphy, C.M.; Haugh, M.G.; O'Brien, F.J. The effect of mean pore size on cell attachment, proliferation and migration in collagen–glycosaminoglycan scaffolds for bone tissue engineering. *Biomaterials* **2010**, *31*, 461–466. [CrossRef] [PubMed]

63. Simske, S.J.; Ayers, R.A.; Bateman, T.A. Porous materials for bone engineering. *Mater. Sci. Forum* **1997**, *250*, 151–182. [CrossRef]

64. Dumas, M.; Terriault, P.; Brailovski, V. Modelling and characterization of a porosity graded lattice structure for additively manufactured biomaterials. *Mater. Des.* **2017**, *121*, 383–392. [CrossRef]

65. Sudarmadji, N.; Tan, J.Y.; Leong, K.F.; Chua, C.K.; Loh, Y.T. Investigation of the mechanical properties and porosity relationships in selective laser-sintered polyhedral for functionally graded scaffolds. *Acta Biomater.* **2011**, *7*, 530–537. [CrossRef] [PubMed]

66. De Wild, M.; Zimmermann, S.; Rüegg, J.; Schumacher, R.; Fleischmann, T.; Ghayor, C.; Weber, F.E. Influence of microarchitecture on osteoconduction and mechanics of porous titanium scaffolds generated by selective laser melting. *3D Print. Addit. Manuf.* **2016**, *3*, 142–151. [CrossRef]

67. Taniguchi, N.; Fujibayashi, S.; Takemoto, M.; Sasaki, K.; Otsuki, B.; Nakamura, T.; Matsushita, T.; Kokubo, T.; Matsuda, S. Effect of pore size on bone ingrowth into porous titanium implants fabricated by additive manufacturing: An in vivo experiment. *Mater. Sci. Eng. C* **2016**, *59*, 690–701. [CrossRef] [PubMed]

68. Fukuda, A.; Takemoto, M.; Saito, T.; Fujibayashi, S.; Neo, M.; Pattanayak, D.K.; Matsushita, T.; Sasaki, K.; Nishida, N.; Kokubo, T.; et al. Osteoinduction of porous Ti implants with a channel structure fabricated by selective laser melting. *Acta Biomater.* **2011**, *7*, 2327–2336. [CrossRef] [PubMed]

69. Khan, W.S.; Rayan, F.; Dhinsa, B.S.; Marsh, D. An osteoconductive, osteoinductive, and osteogenic tissue-engineered product for trauma and orthopedic surgery: How far are we? *Stem Cells Int.* **2012**, *2012*, 236231. [CrossRef] [PubMed]

70. Van der Stok, J.; Van der Jagt, O.P.; Amin Yavari, S.; De Haas, M.F.P.; Waarsing, J.H.; Jahr, H.; Van Lieshout, E.M.M.; Patka, P.; Verhaar, J.A.N.; Zadpoor, A.A.; et al. Selective laser melting-produced porous titanium scaffolds regenerate bone in critical size cortical bone defects. *J. Orthop. Res.* **2013**, *31*, 792–799. [CrossRef] [PubMed]

71. Hollister, S.J. Porous scaffold design for tissue engineering. *Nat. Mater.* **2006**, *5*, 590. [CrossRef]

Article

Experimental Characterization of the Primary Stability of Acetabular Press-Fit Cups with Open-Porous Load-Bearing Structures on the Surface Layer

Volker Weißmann [1,2,*], **Christian Boss** [3], **Christian Schulze** [2], **Harald Hansmann** [1] and **Rainer Bader** [2]

1   Faculty of Engineering, University of Applied Science, Technology, Business and Design, Philipp-Müller-Str. 14, 23966 Wismar, Germany; h.hansmann@ipt-wismar.de
2   Biomechanics and Implant Technology Research Laboratory, Department of Orthopedics, Rostock University Medicine, Doberaner Strasse 142, 18057 Rostock, Germany; christian_schulze@med.uni-rostock.de (C.S.); rainer.bader@med.uni-rostock.de (R.B.)
3   Institute for Polymer Technologies e.V., Alter Holzhafen 19, 23966 Wismar, Germany; boss@ipt-wismar.de
*   Correspondence: weissmann@ipt-wismar.de; Tel.: +49-03841-758-2388; Fax: +49-03841-758-2399

Received: 27 September 2018; Accepted: 16 October 2018; Published: 17 October 2018

Abstract: *Background:* Nowadays, hip cups are being used in a wide range of design versions and in an increasing number of units. Their development is progressing steadily. In contrast to conventional methods of manufacturing acetabular cups, additive methods play an increasingly central role in the development progress. *Method:* A series of eight modified cups were developed on the basis of a standard press-fit cup with a pole flattening and in a reduced version. The surface structures consist of repetitive open-pore load-bearing textural elements aligned right-angled to the cup surface. We used three different types of unit cells (twisted, combined and combined open structures) for constructing of the surface structure. All cups were manufactured using selective laser melting (SLM) of titanium powder (Ti6Al4V). To evaluate the primary stability of the press fit cups in the artificial bone cavity, pull-out and lever-out tests were conducted. All tests were carried out under exact fit conditions. The closed-cell polyurethane (PU) foam, which was used as an artificial bone cavity, was characterized mechanically in order to preempt any potential impact on the test results. *Results and conclusions:* The pull-out forces as well as the lever moments of the examined cups differ significantly depending on the elementary cells used. The best results in pull-out forces and lever-out moments are shown by the press-fit cups with a combined structure. The results for the assessment of primary stability are related to the geometry used (unit cell), the dimensions of the unit cell, and the volume and porosity responsible for the press fit. Corresponding functional relationships could be identified. The findings show that the implementation of reduced cups in a press-fit design makes sense as part of the development work.

Keywords: Ti6Al4V; selective laser melting; mechanical characterization; press-fit; primary stability

---

1. Introduction

Implants today are an important achievement of modern society and an indispensable part of daily life. To improve an implant design, it is important to build a knowledge base that allows insights gained to be integrated into new developments. Modern, generative manufacturing processes provide an excellent foundation for the support and acceleration of the knowledge required in the area of experimental development and for the transfer from result in application [1–4]. Developing implants beyond the current state of the art, for example in the field of orthopedics, is an interesting task for development engineers. Due to their outstanding mechanical and biocompatible properties, titanium and titanium alloys, in addition to other materials, are at the center of development work [5–7].

Of major interest is the implementation of open-porous structures in orthopedic implants. These structural elements provide excellent conditions to fulfil structural and functional requirements. Open-porous structures meet the mechanical requirements regarding surface quality as well as those regarding design conditions [8–10]. In addition, such structures offer a potential for solving the problems of different stiffnesses between human bone and full implants [11,12]. As a result of their geometry, open-pore structures offer the cells good conditions for nutrient supply, and consequently, the possibility to grow well into the pores. Characteristic features of open-pore structures like pore size and distribution as well as connectivity affect biological processes like cell migration and proliferation and as a result the regeneration process [3,13].

The applications of open-porous and load-bearing structures in orthopedic applications range from femoral stems, knee implants to artificial hip cups [3]. Harrison et al. developed a new surface architecture for orthopedic stem components to ensure a greater resistance against transverse motion. This allowed an enhanced primary fixation [14]. Jetté et al. designed a femoral stem with a diamond cubic lattice structure and assessed its potential as a biomimetic construct for load-bearing orthopedic implants [15]. Marin et al. evolved an acetabular cup with Trabecular Titanium^{TM} to increase osseointegration [16].

The design of the area between the implant and human bone or the transition boundary between the implant and human bone is crucial for the success of the substitution of bone with the implant. A large number of investigations are therefore concerned with the implementation of implant surfaces with biocompatible or bioactive properties [17–20]. The aim is to establish conditions that will optimally assist bone in growing in order to achieve maximum secondary stability [21–25].

Numerical simulations are also frequently used in the area of implant development as an indispensable link between constructive development ideas and experimental testing [26–30]. The success of an implantation is determined not only by secondary stiffness but also by primary anchoring strength [29,31,32]. Le Cann et al. investigated the influence of surface roughness on primary stability [33]. Goriainov et al. tested the interaction between the surface properties of the acetabular cup and its initial stability [34]. Gebert et al. studied the influence of press-fit parameters on the primary stability of uncemented femoral head resurfacing prostheses [35]. With this work, an influence of the surface roughness on the primary stability could be demonstrated. It is particularly remarkable that the primary stability can be improved up to a respective roughness value beyond which deterioration occurs is essentially influenced by the cup design. However, the influence of modifications to commercially available implants on primary stability must not be disregarded when considering the entire subject area [36,37]. Primary stability as a prerequisite for good osseointegration significantly influences the success of an implantation [29].

In the field of press-fit cups, experimental work evaluating the pull-out and lever-out behavior in preclinical as well as in post-clinical investigations is of particular interest for the assessment of anchoring strength [38–43]. Besides bones (cadavers) closed-cell foams are being used more and more often in their function as an artificial bone bed [37,44–46]. In addition to different PU (polyurethane) foams, EP-DUR polyurethane foams, polymethacrylamide (PMI) foams and a combination of a polyvinyl chloride (PVC) layer and a PMI foam have served as bone substitutes [47–49]. Although PU foam deviates from the properties of acetabular bone, it is well suited for experimental work due to its uniform cell structure and associated mechanical properties. This is mainly because of the reproducibility of the results, better availability and avoidance of ethical problems.

In the context of this work, standard acetabular cups in the press-fit version were constructively provided with a porous layer on the surface to experimentally determine the influence on primary stability. The porous structures were applied to a reduced-acetabular cup, the suitability of which for the characterization of primary stability has been evaluated in a previous study [50]. All acetabular cups were manufactured using additive manufacturing technology (Selective Laser Melting). The porous surface structures were varied constructively in order to generate different densities in the structural layer and to vary the structure-determining geometry. These constructively produced structures, though differing significantly, nevertheless aim to deliver bone-like properties as a load-bearing structural layer. Thus, forces occurring in the implant bed can be directly absorbed and transmitted by the implant. The porous structure, which has an osteoconductive effect and supports osteoinduction, can significantly improve primary stability [21,25].

The focus of the experimental work is the description of the impact of the applied structural geometry on the primary stability.

2. Materials and Methods

2.1. Cup Design

The modified cups (Figure 1) were designed on the basis of a conventional press-fit cup with a pole flattening. The suitability of a modified press-fit cup (reduced height) for the use in a development phase was verified in an earlier study [50]. All cups were designed in a reduced design with an equatorial cup diameter of 55.3 mm and a pole flattening of 1 mm. The height profiles of the cup were recorded (equatorial cup diameter 55.3 mm; pole flattening 1 mm) by means of a non-contact measuring microscope Mitutoyo—QVE-200 Pro (Mitutoyo Corporation, Kawasaki, Japan), transferred to a CAD model (PTC Creo, Version 3.0, Parametric Technology Corporation, Needham, MA, USA) and redesigned. The pattern used was an Allofit-IT 54/JJ (Zimmer GmbH; Winterthur; Switzerland). The surface structures consist of repetitive open-pore load-bearing textural elements aligned right-angled to the cup surface. The mechanical properties of the selected load-bearing open-pore structure were successfully ascertained in pretests [51–54]. The surface structure was adapted in its outer dimensions to the height profile of the Allofit IT-54/JJ. We have developed three different cup designs with three different types of unit cells (Table 1). Altogether, 8 different press-fit cups have been constructed.

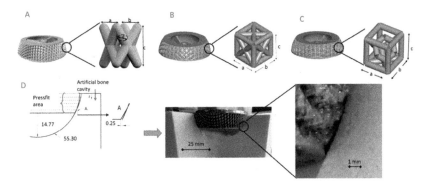

Figure 1. Designs of artificial acetabular cups with an open-porous load-bearing unit cell in a reduced variant; (**A**) Illustration—twisted unit cell, (**B**) Illustration—combined unit cell, (**C**) Illustration—combined open unit cell substitute, (**D**) Press-fit area and gap in case of reduced cup model (negative press-fit)—schematic figure and photograph, all units are in mm.

Cup variant *A* with a twisted unit cell geometry exists in five versions. The unit cells differ in depth *a* between 2.12 mm and 2.83 mm, in width *b* between 2.12 mm and 2.83 mm and in height *c* between 3 mm and 4 mm. The rod diameter *d* varied between 0.8 to 1.1 mm. Cup variant *B* with a combined unit cell geometry exists in two versions. The unit cells have a depth a of 4 mm, width *b* of 4 mm and height *c* of 4 mm. The rod diameter d varied between 0.8 and 0.9 mm. The combined unit cell geometry is designed with a cubic structure with transverse struts on the outer surfaces and a diamond-like structure. Regardless of the force acting on the unit cell, this structure offers very uniform strength. The structure is very suitable for use on the surface of a press-fit cup thanks to its direction-independent nature [54].

Table 1. Overview of the eight different cup-designs, the types of the unit cells (twisted, combined and combined open), the dimensions of the unit cells and porosities and volumes of the press-fit area. All values are derived from CAD data and are given in mm.

Unit Cell	Twisted (V)					Combined (D)		Combined Open (D_o)
Dimension	V4_09	V4_10	V4_11	V3_09	V3_08	D4_09	D4_08	D_o_4_09
Width-*a* (mm)	2.83	2.83	2.83	2.12	2.12	4.00	4.00	4.00
Depth-*b* (mm)	2.83	2.83	2.83	2.12	2.12	4.00	4.00	4.00
Height-*c* (mm)	4.00	4.00	4.00	3.00	3.00	4.00	4.00	4.00
Strut diameter-*d* (mm)	0.90	1.00	1.10	0.90	0.80	0.90	0.80	0.90
Porosity-Structure area (%)	72.50	67.40	60.60	58.80	65.50	61.10	66.90	74.80
Volume-Press-fit area (cm³)	0.30	0.39	0.25	0.32	0.54	0.97	0.91	0.77

Cup variant C with a combined open unit cell geometry exists in one version. The unit cells have a depth *a* of 4 mm, width *b* of 4 mm and height c of 4 mm. The rod diameter *d* is 0.9 mm. The combined unit cell geometry is designed with a cubic and a diamond-like structure without transverse struts on the outer surfaces. Using the overall model of the cups as a basis, reduced designs were created. With the reduction of the acetabular cup, the pole near area was removed, but the press-fit was retained. Cup regions from the press-fit regions protrude so far that a gap of 0.25 mm is created between the artificial bone bed and the cup (negative press-fit-Figure 1-Area D).

The following expression was used to calculate the porosity of load-bearing structure volume from the CAD data:

$$\text{Porosity} - \text{structure area} = \left(1 - \frac{V_{str}}{V_{full}}\right) \cdot 100\% \tag{1}$$

where V_{str} is the volume of the area with the struts and V_{full} is the overall volume of this area in a closed manner.

The volume (Press-fit area) produced by the structured section of the cups was also calculated by CAD. For this intention, it was virtually determined how large the volume is that penetrates the artificial bone cavity (Figure 2). The acetabulum and artificial bone cavity were positioned in the CAD system in the same way as in the test situation. The results for every cup-design are given in Table 1.

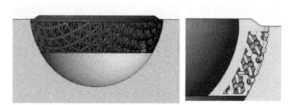

Figure 2. The cup is positioned in the artificial bone cavity (**left**) and the area virtually penetrates the artificial bone cavity-red hatched area (**right**). This area describes the Press-fit volume.

2.2. Fabrication

(1) The acetabular cups considered in this paper were manufactured by C. F. K. CNC-Fertigungstechnik Kriftel GmbH (C. F. K. CNC-Fertigungstechnik Kriftel GmbH, Kriftel, Germany) using selective laser melting with a SLM 280. Titanium powder (Ti6Al4V) with a mean particle size of 43.5 µm was used for their manufacture in a highly pure argon atmosphere. All parts were built using identical processing parameters (Table 2) in the same orientation and on a substrate plate with a support structure. The support structures were removed mechanically by hand.

Table 2. SLM process-energy-relevant process parameters.

Parameter	Description	Unit	Process Parameter
P	Laser power	W	275
v	Scan speed	mm/s	805
d	Hatch spacing	µm	120
t	Layer thickness	µm	50

(2) For the production of artificial bone cavities Sika Block M 330 (Sika GmbH, Stuttgart, Germany) was applied. This material, a thermosetting polyurethane with closed cells, is ideally suited for a comparative evaluation of the relevant acetabular cups. The properties comprise from a density of 0.24 g/cm^3 (according to test standard ISO 845) and a compressive strength of 4 MPa (according to test standard ISO 844) to an elastic modulus of 150 MPa (according to test standard ISO 850).

The material was provided in plate form in the dimensions 1000 × 500 mm. The artificial bone cavities were manufactured using a CNC milling machine i-mes-FLATCOM 50-VH (i-mes GmbH, Eiterfeld, Germany) using the plate.

The artificial bone cavities were manufactured as described in Weißmann et al. Since the mechanical properties of the plate vary across the width of the plate due to the manufacturing process, the cavities were used for each acetabulum from a corresponding material line ($n = 5$) [50].

2.3. Measurements

The measurements of the following points were carried out extensively as described in Weißmann et al. [50]. Here, the relevant points are briefly explained.

(1) The measurements of the acetabular cups as well as the artificial bone cavities, both being relevant for the press-fit, were performed with a non-contact measuring microscope (Mitutoyo-QVE-200 Pro; Mitutoyo Corporation, Kawasaki, Japan). Based on the measurement points, circles of best fit were determined using the method of least squares. The outlier identification and elimination from the measurement data due to light reflections and loose PUR particles was performed using a box plot (according to John W. Tukey) in a Matlab script. To verify the actual press-fits and for quality control, the resulting replacement diameters were used.

(2) In all cases, the assessment of the primary stability (anchoring strength) of the press-fit cups was realized by pull-out tests (Figure 3) with a universal testing machine (INSTRON E 10,000; Instron GmbH, Darmstadt, Germany). The cups were first press-fitted into the artificial bone cavities until they were flush with the edge of the cavity. Following this, the cups were pulled out of the cavity using a pull-out stamp. The speed for both the press-fit of the cup into the bone cavity and the pull-out of the cups was 5 mm/min. In the measurements, each performed 5 times per press-fit cup, the effective measurement data ($F_{pull-out}$) were recorded. As primary pull-out stability the first force maximum was used.

(3) The assessment of the initial tangential stability of the acetabular cups were realized by lever-out tests (Figure 4) with a universal testing machine (Zwick Z50; Zwick GmbH & Co. KG, Ulm, Germany). The cup was first pressed into the artificial bone cavity until the edge of the cup is flush with the bone bed. The cup was first pressed into the artificial bone cavity until the edge of the cup was flush with the bone bed. The cup was then vertically loaded with a force until it was released. The

first local maximum (F_L) load was evaluated as the primary lever-out stability, which at the same time indicates the beginning of the movement of the cup in the bone cavity. The speed for the press-fit of the cups into the bone cavity and the lever-out of the cups was 5 mm/min. A moment M_I of 0.62 Nm, resulting from the dead weight (0.87 kg) and length (178.3 mm) of the lever, was also integrated into the calculation.

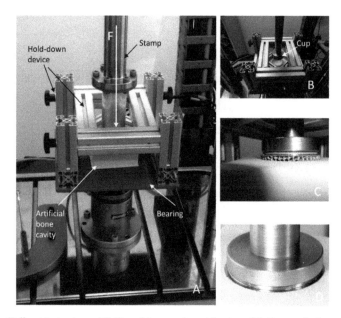

Figure 3. Pull-out-test setup—(**A**) Complete experimental setup; (**B**) Cup ready for pressing in; (**C**) View from upside of the acetabular cup with artificial bone cavity and the pull-out stamp; (**D**) Cup completely press-fitted.

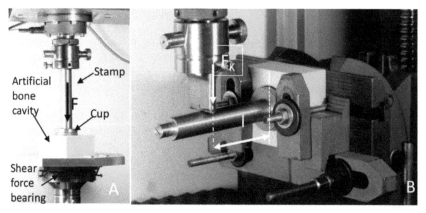

Figure 4. Pull-out-test setup—(**A**) Experimental setup-press-fitting; (**B**) Experimental setup-levering out.

The lever-out moment was calculated as follows:

$$M_L = F_L \cdot l + M_I \tag{2}$$

In the calculation is F_L the maximum lever-out tilting force, l the lever length and M_I the specific moment.

On the basis of the determined force F_L and the displacement of the cup in the bone cavity, it is possible to evaluate the work required to lever out the cup.

The lever-out work was calculated as

$$W = F_L \cdot s \tag{3}$$

from the lever-out tilting force F_L and the displacement s of the cup.

2.4. Statistical Analysis

All data listed in tables are expressed as mean values $\pm$ standard deviation (SD). A non-linear regression with Excel 2016 for Windows was used to display the relationships between the volume of the press-fit area and the lever-out moment as well as the pull-out force.

All statistical analyses were made using SPSS, software version 22 for Windows (SPSS® Inc. Chicago, IL, USA). For the pull-out force, the lever-out moment and the lever-out work, a one-way ANOVA followed by Dunn's T3 post-hoc test was made to statistically examine significant differences between the means. The results from this comparison were shown in a boxplot. A significance level of $p < 0.05$ was regarded as statistically significant.

3. Results and Discussion

3.1. Accuracy of Fabricated Samples

Table 3 lists the dimensions determined for the artificial bone cavity and the acetabular cups. The press-fit of the cups are calculated as the difference between the best fit circle of the press-fit cups and the best-fit circle of the artificial bone cavity

Table 3. Accuracy of fabricated bone cavities (diameter cavity) and acetabular cups (equatorial diameter) as well as the resulting press-fits of these combinations. The values from the bone cavities are given as the arithmetical average ($n = 5$).

Name	Press-Fit Cup		Artificial Bone Cavity		Press-Fit (mm)
	Best Fit Circle (mm)	Roundness (mm)	Best Fit Circle (mm)	Roundness (mm)	
V3_08	55.32	0.26	53.18 ± 0.02	0.14 ± 0.02	2.13 ± 0.02
V3_09	55.47	0.17	53.34 ± 0.02	0.13 ± 0.04	2.13 ± 0.01
V4_09	54.90	0.29	52.68 ± 0.01	0.15 ± 0.01	2.16 ± 0.01
V4_10	55.03	0.02	52.87 ± 0.01	0.14 ± 0.02	2.15 ± 0.01
V4_11	55.20	0.28	53.07 ± 0.01	0.12 ± 0.01	2.13 ± 0.01
D4_08	54.98	0.30	52.87 ± 0.01	0.14 ± 0.01	2.11 ± 0.01
D4_09	55.04	0.11	52.87 ± 0.01	0.14 ± 0.01	2.17 ± 0.01
D_o_4_09	55.03	0.25	52.87 ± 0.01	0.14 ± 0.01	2.16 ± 0.01

The processing values for the artificial bone cavities were determined based on the values for press-fit cups. The aim was to provide a constructive press-fit of 2 mm for all cup-bone cavity pairs.

For all pairings a press-fit was achieved between a minimum of 2.11 mm and a maximum of 2.17 mm. The deviations among each other amount to a maximum of 0.06 mm. With respect to the minimum possible press-fit, this is less than 3% (2.84%). The roundness values of the bone cavity of 0.12 to 0.15 demonstrate the high repeatability of the manufacturing method for artificial bone cavities. The roundness values of the press-fit cups from 0.02 to 0.30 vary slightly more. With respect to the additive manufacturing process, these are excellent results [55–58].

Dimensional deviations or differences in the produced press-fit can lead to different insertion forces. These differences would be the cause of stress differences in the bone cavity and unequal conditions for the contact of the press-fit cup with the surface of the bone cavity. The resulting deviations produce differences in tension in the bone cavity and create different conditions for the movements of the press-fit cup in the bone cavity [44,49]. Only if the conditions for the generation of a good primary stability are given, can corresponding good long-term results be expected [27].

Overall, it can be assumed that the differences between each other are so small that this will have no effect on the assessment of the primary stability of the artificial acetabular cups. The press-fit results are only so slightly different that the results in the pull-out test and the lever-out test are not affected.

3.2. Pull-Out Force

To determine the pull-out forces, the manufactured cups were stripped from the cavities after being press-fitted into the artificial bone cavity. The results are shown in Figure 5 and Table 4.

Figure 5. Boxplots of the measured pull-out force (N). Boxplots indicate the median value, the interquartile range (IQR: interval between the 25th and 75th percentile, blue rectangle) and the extremum values ($n = 5$).

Table 4. Significances of the determined pull-out-forces from the different press-fit cups. For statistical analysis one-way ANOVA with Dunn's T3 post-hoc test was conducted. Values of $p < 0.05$ were set to be significant (N.S.—not significant).

Cupversion	D4_09	D_o_4_09	V3_08	V3_09	V4_09	V4_10	V4_11
D4_08	N.S.	0.00438	<0.001	<0.001	<0.001	<0.001	<0.001
D4_09	-	0.00193	<0.001	<0.001	<0.001	<0.001	<0.001
D_o_4_09	-	-	<0.001	0.0006	<0.001	<0.001	<0.001
V3_08	-	-	-	N.S.	N.S.	0.0242	N.S.
V3_09	-	-	-	-	N.S.	N.S.	N.S.
V4_09	-	-	-	-	-	N.S.	N.S.
V4_10	-	-	-	-	-	-	N.S.

The results of the experiments carried out according to the measuring methodology reveal differences that are related to the structural elements used. Whereas the combined structures achieve the highest results (D4_08 = 708 N; D4_09 = 704 N), the pull-out forces for the twisted structures (Max: V3_08 = 351 N; Min: V4_10 = 308 N) are significantly lower. The combined open structure (550 N) lies between the two combined variants and the cups with the twisted structures.

After carrying out a statistical significance test using one-way Anova with Dunnett's T3 post-hoc test (multiple comparisons), the following relationships become clear. The two combined structures do not differ significantly from each other. However, the combined open structure is significantly below the combined structure (D4_08 to D_o_4_09/p = 0.00438; D4_09 to D_o_4_09/p = 0.00193). The differences in the twisted structures are consistently significant (values see Table 4). In the twisted structures only version V3_08 deviates significantly from version V4_10 (p = 0.0242). The differences between the combined open and twisted structures can mainly be explained by the existing differences in press-fit volume. The press-fit volumes of the combined (D4_08 = 0.91 cm^3; D4_09 = 0.97 cm^3) and the combined open structure with 0.77 cm³ clearly differ from the twisting structures (<0.54 cm³). However, this relationship is not identifiable in the twisting structures, since despite clear differences in the press-fit volume between the twisting structures, a significant difference could only be determined between the variants V3_08 and V4_10. It seems that in addition to the press-fit volume, other influencing factors such as the surface quality (roughness and manufacturing accuracy) of the struts of the structure and their dimensions (length, diameter, surface area) could play a role [55,59].

The pull-out behavior of the different cup models is shown in Figure 6. The representation of the force profiles over cup displacement in the artificial bone cavity additionally offers the possibility to evaluate the measured maximum force in relation to the reached cup displacement at that time. The curves show characteristic differences.

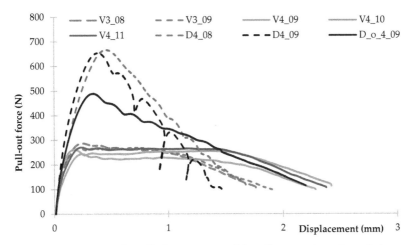

Figure 6. Representative force-displacement curve of the pull-out tests for each cup design.

The curve for the cups with a combined structure differs clearly from the curves for the cups with a combined open or twisting structure. The most striking feature here is the cascading force decrease after a maximum force has been exceeded. This cascade is characterized in that a renewed force increase is determined after a drop in force. This course reflects the loosening and re-jamming of the cup in the artificial bone cavity. These cascades are most pronounced in version D4_09. This cascade development is also evident in the combined open structure version D4_08, though weaker. Apparently, this cascade is due to the larger space between the individual struts or the greater porosity. Here, the material of the artificial bone cavity has the possibility to fill more space. The necessary release from this room requires force again.

This cascade is characterized in that a renewed force increase is determined after a drop in force. This course reflects the loosening and re-jamming of the cup in the artificial bone cavity. These cascades are most pronounced in version D4_09. This cascade development is also evident in the combined open structure version D4_08, though weaker. Apparently, this cascade is due to the larger space between the individual struts or the greater porosity. Here, the material of the artificial bone cavity has the possibility to fill more space. The necessary release from this room requires force again.

The number of cascades obviously results from the number of superficial, continuous struts (Figure 7—red lines). The maximum peak (and thus the first peak of force) results from overcoming the edge of the hip cup. The second to fifth peak results from the strut contours. Starting at the highest point of the continuous strut lines. The differences in cascade intensity of the cup variants are caused by the differences in the strut diameter. The strut with a rod diameter of 0.9 mm has a larger contact surface to the artificial bone bed. This requires more force to loosen from the artificial bone cavity. The differences between the open and closed variants (D_o_4_09 and D4_09) are due to the varying degrees of free space in the surface of the hip cups. More free space (D_o_4_09) requires less force than with the closed variant (D4_09).

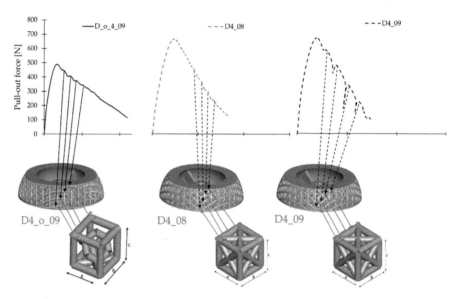

Figure 7. Representation of the cascades with reference to the structure on the cup surface.

The press-fit cups with the twisted structure show a completely different behavior. After reaching the force maximum, the corresponding force path continues at a uniform level of force. This applies to the twisted structure with a height of 3 mm as well as to the structure with a height of 4 mm. It is clearly shown here; however, that the versions in the 4 mm height maintain this level of force significantly longer. A weakening of the cup anchoring takes place here only after about 1.5 mm compared to about 1 mm in the variants with a height of 3 mm. Here, the cups with the structural elements whose individual elements have a height of 4 mm and an associated spacing of the bars of 2.83 mm, provide the artificial bone cavity material more space for anchoring than the variant of 3 mm height and a spacing of 2.12 mm. As a result, the force is maintained longer at one level.

In view of later desired ingrowth of the bone into the structural area as well as the formation of blood vessels, larger open areas have advantages over the smaller areas [22,25,60]. Here it is important to carefully observe the interaction of the geometric conditions (unit cell and macro-porosity) and the component properties influenced by the additive manufacturing process (e.g., roughness or micro-porosity, surface finish at intersections) [61–63].

While the diamond structures reach the maximum force required to pull out at approx. 0.6 to 0.7 mm, these values for the twisted structures are approx. 0.2 to 0.3 mm. The open combined structure shows a maximum at approx. 0.35 mm. In addition, it can be seen that the twisted version with a height of 3 mm as well as the combined structure D4_08 still require approximately 100 N after about 1.6 to 1.8 mm displacement for a further release.

In the case of the twisted versions with a height of 4 mm and the open combined structure, the cups have already experienced a displacement of approximately 2.5 mm at a force of 100 N. The progression curves of the press-fit cups are very similar. This value probably reflects the interaction between the artificial bone cavity and the surface of the additively manufactured cup.

As can be seen from Figure 8, all cups leave clear traces of an impression on the entire circumference of the artificial bone bed. The evaluation of these traces using this visual assessment of the contact surface has been described, for example, by Le Cann et al. to characterize how the roughness of a cup affects primary stability [33].

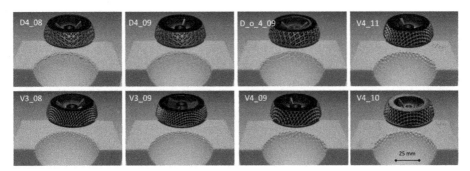

Figure 8. Representative pictures of the bone cavities after the pull-out test for each cup design.

All cups left distinct positioning traces in the press-fit region. The artificial bone cavity remained intact. The artificial bone cavities shown in Figure 8 exhibit clear marks of an anchorage. The damage patterns of the artificial bone cavity differ optically from each other.

All twisted versions show dot-like impressions in the cavity area. The cavity edges remain sharply intact. Differences caused by the different bar diameters (3 and 4 mm) and bar distances (2.83 and 2.12 mm) are optically present. With increasing bar diameter, the damage in the bone bed also increases. Variant V4_11 shows clearer and stronger traces than versions V4_10, V4_09, V3_08 and V3_09.

The combined structures (D4_08, D4_09) show rather flat impressions on the artificial bone cavity areas. The cavity edges tend to blur slightly, as a representation of slight material detachments. These detachments are much less pronounced in the diamond open structure.

The forces determined in the pull-out test and the traces in the bone bearing also allow the following conclusion to be drawn. The twisting structure already destroys the corresponding area in the bone bearing during the press fitting. Because of that, less force is required when pulling out of the bearing because the resistances against loosening are lower than with intact material. The combined structure, on the other hand, only damages the bone bearing when it is pulled out. Here, the resistance of predominantly intact material must be overcome. This leads to a higher power requirement.

In addition, the contacting of the structures with the bone bed takes place differently. The contact of the twisting structure is made punctually. The combined and combined open structure creates a two-dimensional contact to the surface of the bone bed. To overcome the press fit, more force is required for the two-dimensional contacts than for the punctual contacts.

3.3. Lever-Out Moment

After being press-fitted into the artificial bone cavities, all cup models were levered out from the cavities to determine the lever-out moments as described in 0. The results are shown in Figure 9 and Table 5. The course of the forces required to lever out the cups over the displacement is shown in Figure 10.

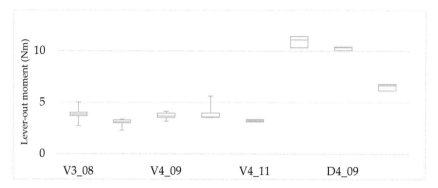

Figure 9. Boxplots of the measured lever-out moments (Nm). Boxplots indicate the median value, the interquartile range (IQR: interval between the 25th and 75th percentile, blue rectangle) and the extremum values ($n = 5$).

Table 5. Significances of the determined lever-out-moments for the different press-fit cups. For statistical analysis one-way ANOVA with Dunn's T3 post-hoc test was conducted. Values of $p < 0.05$ were set to be significant (N.S.—not significant).

Cupversion	D4_09	D_o_4_09	V3_08	V3_09	V4_09	V4_10	V4_11
D4_08	N.S.	<0.001	<0.001	<0.001	<0.001	<0.001	<0.001
D4_09	-	<0.001	<0.001	<0.001	<0.001	<0.001	<0.001
D_o_4_09	-	-	<0.001	<0.001	<0.001	<0.001	<0.001
V3_08	-	-	-	0.04619	N.S.	N.S.	0.04649
V3_09	-	-	-	-	N.S.	N.S.	N.S.
V4_09	-	-	-	-	-	N.S.	N.S.
V4_10	-	-	-	-	-	-	N.S.

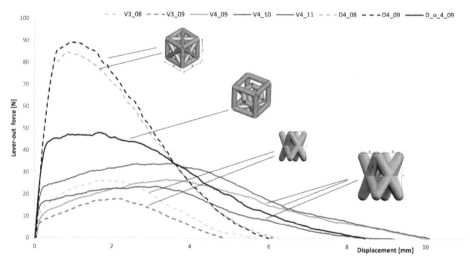

Figure 10. Representative lever-out force vs. displacement curve of the lever-out test for each cup design.

The influence of the applied structural elements on the behavior of the press-fit cups in the lever-out test can be clearly established on the basis of the experimentally determined lever-out moments. The best results were achieved by the combined structure (D4_08 = 10.9 Nm,

D4_09 = 10.3 Nm), followed by the combined open structure (6.5 Nm) and the twisted structure (Max: V3_08 = 3.9 Nm; Min: V3_09 = 3.1 Nm).

By carrying out a statistical significance test using one-way Anova with Dunnett's T3 post-hoc test (multiple comparisons) it is possible to describe the following relationships. The two combined structures do not differ significantly from each other. However, the combined open structure is significantly below the combined structure (D4_08 and D4_09 to D_o_4_09/$p < 0.001$).

The differences of the experimentally determined lever-out moments shown between the combined structures, the combined open structures and the twisted structures are significant in all cases ($p < 0.001$). For the twisted structures, only the version V3_08 deviates significantly from both version V3_09 ($p = 0.04619$) and version V4_11 ($p = 0.04649$). Similar to the pull-out tests, the differences between the combined and the combined open structures to the twisted structures can be explained by the existing differences in press-fit volume. The differences between the structure V3_08 and V3_09 and V4_11 also result from the differences in the press-fit volumes (V3_08 = 0.54 cm^3; V3_09 = 0.32 cm^3; V4_11 = 0.25 cm^3). The fact that variant V4_09 does not deviate significantly from variant V3_8 despite a lower press-fit volume (0.3 cm^3) is additional evidence that other factors are notoriously influencing the anchoring strength.

The lever-out behavior of the tested cup models is shown in Figure 10. All additively manufactured cups show curves which are characteristic for the structural elements used.

All models were preloaded with an initial moment of 0.62 Nm by the self-weight of the test setup. The representation of lever-out forces over displacement displays for the combined structure a maximum lever-out force (mean values: D4_08 = 90.3 N; D4_09 = 85 N) at a displacement of approx. 1 mm and then a decrease of the moment up to a displacement of 6 mm. The combined open structure reaches a lever-out force maximum (mean value: 51.6 N) after approx. 1.8 mm. This cup variant reduces the force to zero after a displacement of about 8.3 mm. The twisted structures show differences depending on the size of the structure. The twisted structures with dimensions of 3 mm height reach a lever-out force maximum (mean values: V3_08 = 29 N; V3_09 = 22 N) after about 1.8 to 2.2 mm. The twisted structures with dimensions of 4 mm height reach force maximums (mean values: V4_09 = 27.1 N; V4_10 = 27.5 N; V4_11 = 23.2 N) after about 3.5 to 3.7 mm. The force reduction continues in the V3-versions up to a displacement of approx. 4.8 to 5.8 mm. The V4 versions run to zero at about 9 to 10.5 mm.

Similar to the pull-out tests, it can be seen that, following a steep rise, the cups with the combined structure show a continuous force drop after reaching a lever-out force maximum. The combined open structure and the twisted structures behave differently. Here the maximum force is only reached after passing through a plateau phase. This plateau phase is much longer for the V4-variants than for the V3-variants.

This functional difference is related to the geometric design of the individual structures. As shown in Table 1, the combined structures are structures that produce a relatively uniformly shaped surface whose interstices engage only weakly in the bone bed. Here the press-fit is in the foreground.

In the combined open structure and the twisted structures, the shaped surface of the cups is much more open. These structures engage more clearly in the artificial bone stock. The differences between the V3 and V4 variants are due to the geometric dimensions of the individual rods. The larger-sized rods of the V4 variant have larger gaps than the V3-variants (V4-2.83 mm and V3-2.12 mm). Thus, a hooking of the structural elements in the bone cavity in the V4-variant is possible across a longer distance than in the V3 variant.

This leads to differences in the height of the moments determined due to the structure design. In addition, it becomes clear that the twisted structures in the artificial bone bed produce deeper punctual impressions. During the lever-out test, the struts move along these impressions. This behavior is recognizable for all twisted structure variants by traces between the punctual impressions. The illustrations of the bone beds after the pull-out test (Figure 8) do not show these traces. Therefore, due to the already damaged surface, less force is required to lever-out. The twisted structures thereby show overall lower moments than the combined and combined open structure due to the different nature of the unit cell.

A larger structural design is helpful in terms of the positive effects for bone ingrowth [25]. In addition to good primary stability, the bone-like properties of the load-bearing structural layer are an essential prerequisite for good secondary stability of the implant [64]. Secondary stability is essentially characterized by the ability of bone to grow onto the implant surface and thereby firmly anchor the implant. The use of open-pore structures enlarges the implant surface and thus improves the prerequisite for the formation of sufficiently high secondary stability. In addition, a high primary anchoring strength is the prerequisite for creating a sufficiently high secondary stiffness, since only then is sufficient growth of the bone on the surface possible. Only if a load transfer via the implant into the surrounding bone is possible without stress-shielding can a successful use of the implants be ensured. With regard to the geometric selection of structural elements, this circumstance must be taken into account [65]. The combined structures, which are more direction-independent in their properties, show slight advantages here [52,66].

The artificial bone cavities show distinct traces left by the lever-out of the cups. In the following Figure 11 the cup models are shown with representative examples of the artificial bone cavity. The artificial bone cavity is intact despite clear traces of anchoring. The damage patterns of the bone cavities differ optically from each other, as in the case of the pull-out experiments. The twisted versions show, as expected, punctually impressions in the bone cavities. The edge of the cavity remains sharp. The different strut diameters and spaces of the struts in the structure produce visually recognizable representative patterns (dot-like impressions). The combined structures leave flat traces on the bone cavities. The edge of the bone cavities tends to blur slightly, as a representation of slight material detachments. These detachments are significantly less pronounced in the combined open structure.

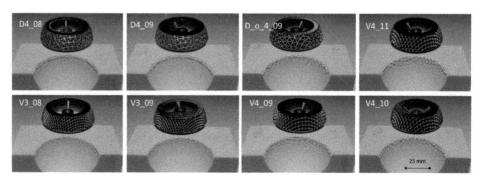

Figure 11. Representative imaging of mechanical deformations in the artificial bone cavity for cup design usage in lever-out test.

The use of an artificial bone cavity has a positive effect on the characterization of primary stability. This speaks in favor of the experimental results determined here since possible property variations, as they occur in the use of cadaveric models, have been omitted. Goldman et al. compared the effect of component surface roughness at the bone implant interface and the quality of the bone on initial press-fit stability [67]. They found no significant differences between the bending moment at 150 m for two kind of press-fit cups with different coefficients of friction. They made clear in the discussion that the results from the use of the cadaveric models represent a realistic representation of surgical interventions, but are also associated with corresponding scatter of the results. For the purpose of this study, which is to evaluate structurally differently designed press-fit cups, the artificial bone bed is the better choice. The uniform mechanical properties of the artificial bone bed provide a much better basis for a comparative consideration of the different cup designs.

3.4. Lever-Out Momentmechanical Work

The lever-out work shown in Figure 12 illustrates the individual force differences required to loosen the cups from the artificial bone cavities. The moment of relaxation thus represents the beginning of the failure.

Figure 12. Boxplots of the measured mechanical work (Nmm) during the lever-out test. Boxplots indicate the median value, the interquartile range (IQR: interval between the 25th and 75th percentile, blue rectangle) and the extremum values ($n = 5$).

The best results were achieved with the cup versions V4_10 (69.9 Nmm), D4_08 (68.8 Nmm) and V4_09 (66.9 Nmm), followed by versions V4_11 (54.8 Nmm), D4_09 (53.1 Nmm) and D_o_4_09 (52.5 Nmm). Much less work was afforded for the loosening of versions V3_08 (48 Nmm) and V3_09 (32 Nmm).

After carrying out a statistical significance test (results Table 6) using one-way Anova with Dunnett's T3 post-hoc test (multiple comparisons), the following coherences become clear. The combined structures D4_08 ($p = 0.04296$) and D4_09 ($p = 0.01733$) deviate significantly from version V3_09. The twisted structure V3_09 deviates significantly from versions V4_09 ($p = 0.01595$), V4_10 ($p = 0.03089$) and V4_11 ($p = 0.01335$).

Table 6. Significances of the determined lever-out work for the different press-fit cups. For statistical analysis one-way ANOVA with Dunn's T3 post-hoc test was conducted. Values of $p < 0.05$ were set to be significant (N.S.—not significant).

Cupversion	D4_09	D_o_4_09	V3_08	V3_09	V4_09	V4_10	V4_11
D4_08	N.S.	N.S.	N.S.	0.04296	N.S.	N.S.	N.S.
D4_09	-	N.S.	N.S.	0.01733	N.S.	N.S.	N.S.
D_o_4_09	-	-	N.S.	N.S.	N.S.	N.S.	N.S.
V3_08	-	-	-	N.S.	N.S.	N.S.	N.S.
V3_09	-	-	-	-	0.01595	0.03089	0.01335
V4_09	-	-	-	-	-	N.S.	N.S.
V4_10	-	-	-	-	-	-	N.S.

In the pull-out test (determined force) and lever-out test (determined moment), the twisted structures perform worse in the evaluation than the combined and combined open structure. However, in the mechanical work determined, the twisted structures with a strut diameter of 4 mm achieve equivalent results here. One reason seems to be that the struts pressed into the artificial bone bed material move along the entire lever-out process in the bone bed material. This means that permanent work has to be done to move the cup further out of the bone bed. This is clearly demonstrated by the curves in Figure 10. This is also supported by the fact that the variants V4_10 and V4_09 achieve the highest values in the work determined. These variants also have the largest gaps between the struts, followed by version V4_11. A larger gap also has a higher proportion of material in the gap than smaller gaps. More material at the same time means more work to overcome the resistance. In total, this means that the work performed for the cup variants D4_08 and V4_09 and V4_10 is comparable. This fact is supported by the results of the cup variant V3_09, which has the lowest porosity (58.8 %) compared to all the other variants tested.

3.5. Correlations—Lever-Out Moment and Pull-Out Force Versus Volume of the Press-Fit Area

Anchoring strength is significantly influenced by the structure used, with its open-porous design characterizing the area that represents the press-fit. When looking at the volume characteristic of each cup variant in relation to the pull-out force or the lever-out moment (Figure 13), it can be seen that the pull-out force and the lever-out moment could be determined by a direct functional relationship, which can be described using a non-linear regression. An exponential function was found which describes the results of the experimental investigations very well. The curve clearly shows that the pull-out forces as well as the lever-out moments are relatively uniform up to a press-fit volume of 0.39 cm^3, followed by a strong increase towards higher press-fit volumes. At high volumes (>0.9 cm^3), the results are very similar for the pull-out forces as well as for the lever-out moments.

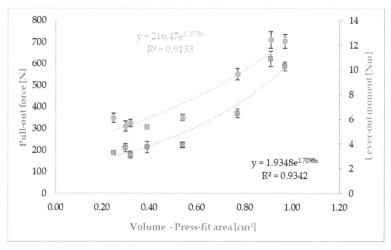

Figure 13. Pull-out force as calculated from pull-out testing and volume press-fit area as well as lever-out moment as calculated from lever-out testing for the eight cup-designs. Results are shown as mean values with the corresponding standard deviation (*n* = 5 for each design).

Both dimensions show an exponential functional relation to the press-fit volume, which is reflected by strong regression coefficients (R^2 = 0.9342 for pull-out force, R^2 = 0.9133 for lever-out moment). This makes it clear that an increase in anchoring strength can be achieved with increasing press-fit volume.

Although the press-fit volume used for this reference does not represent the full volume that actually penetrates the area of the artificial bone, it does directly represent the volume that creates the press-fit.

The determined functional relationships as well as the experimentally obtained measurement results provide a good basis for the selection of appropriate structural elements for the final development of press-fit acetabular cups, which ensure an increase in primary anchoring strength. In particular, the geometric design of the structural elements can thus be used in a targeted manner in conjunction with the mechanical properties and porosity [52,66,68–70]. Also, the determined functional relationships prove that the influence of the volume responsible for the actual press-fit is significantly greater than the porosity. However, since porosity is a measure relevant to secondary anchoring strength, it must not be disregarded.

Structured press-fit cups present an interesting solution, especially with regard to strong pelvic defects (D'Antonio type II). Due to the geometric freedom in structure design and possible size variations, these types of press-fit cups could offer advantages over non-structured cups in anchoring strength [71].

The characterization of the structurally differently designed press-fit cups with two test methods as well as the evaluation of the results in relation to different influencing factors makes a distinctive estimation of the types of cups possible. While the evaluation of anchoring strength with only one procedure or from one aspect is being discussed controversially, a good summary can be made in this study [67]. Several factors, such as material and surface structure (e.g., bead or wire) have been shown to be responsible for bone ingrowth [72]. The press-fit cups used here in this study have almost identical properties so that these can be neglected in the consideration.

The characterization of the cup variants based on the experimentally determined results offers the possibility to capture significant influences and thus show differences. The functional relationships also offer the opportunity to actively intervene in the constructive process and influence the structure design based on the results.

4. Conclusions

In this study, acetabular press-fit cups with a porous, load-bearing structural layer were examined for primary stability. The press-fit cup used was a design developed and evaluated in a previous study.

The porous, load-bearing structural layer was formed from geometrically differently designed unit cells. The preparation was carried out by means of selective laser melting of TiAl6V4. As an artificial bone cavity a PU foam was used, which was characterized experimentally in terms of mechanical properties.

The results show significant differences in the experimentally determined pull-out force, lever-out moment and lever-out-work results. The best results in pull-out and lever-out moments are achieved by the press-fit cups made in the combined structure (denoted D4_08 and D4_09). When looking at the work required to lever out the press-fit cups, it is noticeable that the press-fit cups designated as D4_08, V4_09 and V4_10 achieved the best results.

Overall, it becomes clear that the results for the evaluation of primary stability are related to the geometry used (unit cell), the dimensions of the unit cell, and the volume and porosity which are responsible for the press fit. Corresponding functional relationships could be determined.

The results of the work provide an excellent starting point for the development of press-fit acetabular cups with increased primary stability as a basis for high secondary stability.

Author Contributions: We point out that all authors were fully involved in the study and in preparing the manuscript. V.W. and C.S. designed the study. V.W. generated the CAD samples with support of C.B. and was involved in the manufacturing process of the scaffolds. V.W. and C.S. performed the experiments, analyzed the data with support of C.B. and wrote the initial manuscript. H.H. and R.B. organized the research funding. All authors ensured the accuracy of the data and the analyses and reviewed the manuscript in its current state.

Funding: This research was funded by Federal Ministry of Education and Research grant number 03FH005IX5.

Conflicts of Interest: The authors declare no conflict of interest.

References

1. Harrison, N.; Field, J.R.; Quondamatteo, F.; Curtin, W.; McHugh, P.E.; Mc Donnell, P. Preclinical trial of a novel surface architecture for improved primary fixation of cementless orthopaedic implants. *Clin. Biomech.* **2014**, *29*, 861–868. [CrossRef] [PubMed]
2. Levine, B. A new era in porous metals: Applications in orthopaedics. *Adv. Eng. Mater.* **2008**, *10*, 788–792. [CrossRef]
3. Murr, L.E. Open-cellular metal implant design and fabrication for biomechanical compatibility with bone using electron beam melting. *J. Mech. Behav. Biomed. Mater.* **2017**, *76*, 164–177. [CrossRef] [PubMed]

4. Sing, S.L.; An, J.; Yeong, W.Y.; Wiria, F.E. Laser and electron-beam powder-bed additive manufacturing of metallic implants: A review on processes, materials and designs. *J. Orthop. Res.* **2016**, *34*, 369–385. [CrossRef] [PubMed]

5. Geetha, M.; Singh, A.K.; Asokamani, R.; Gogia, A.K. Ti based biomaterials, the ultimate choice for orthopaedic implants—A review. *Prog. Mater. Sci.* **2009**, *54*, 397–425. [CrossRef]

6. Tan, X.P.; Tan, Y.J.; Chow, C.S.L.; Tor, S.B.; Yeong, W.Y. Metallic powder-bed based 3D printing of cellular scaffolds for orthopaedic implants: A state-of-the-art review on manufacturing, topological design, mechanical properties and biocompatibility. *Mater. Sci. Eng. C* **2017**, *76*, 1328–1343. [CrossRef] [PubMed]

7. Schulze, C.; Weinmann, M.; Schweigel, C.; Keßler, O.; Bader, R. Mechanical Properties of a Newly Additive Manufactured Implant Material Based on Ti-42Nb. *Materials* **2018**, *11*, 124. [CrossRef] [PubMed]

8. Murr, L.E.; Amato, K.N.; Li, S.J.; Tian, Y.X.; Cheng, X.Y.; Gaytan, S.M.; Martinez, E.; Shindo, P.W.; Medina, F.; Wicker, R.B. Microstructure and mechanical properties of open-cellular biomaterials prototypes for total knee replacement implants fabricated by electron beam melting. *J. Mech. Behav. Biomed. Mater.* **2011**, *4*, 1396–1411. [CrossRef] [PubMed]

9. Do Prado, R.F.; De Oliveira, F.S.; Nascimento, R.D.; De Vasconcellos, L.M.R.; Carvalho, Y.R.; Cairo, C.A.A. Osteoblast response to porous titanium and biomimetic surface: In vitro analysis. *Mater. Sci. Eng. C* **2015**, *52*, 194–203. [CrossRef] [PubMed]

10. Wang, X.; Zhou, S.; Xu, W.; Leary, M.; Choong, P.; Qian, M.; Brandt, M.; Xie, Y.M.; Xu, S. Topological design and additive manufacturing of porous metals for bone scaffolds and orthopaedic implants: A review. *Biomaterials* **2016**, *83*, 14. [CrossRef] [PubMed]

11. Limmahakhun, S.; Oloyede, A.; Sitthiseripratip, K.; Xiao, Y.; Yan, C. Stiffness and strength tailoring of cobalt chromium graded cellular structures for stress-shielding reduction. *Mater. Des.* **2017**, *114*, 633–641. [CrossRef]

12. Simoneau, C.; Terriault, P.; Jetté, B.; Dumas, M.; Brailovski, V. Development of a porous metallic femoral stem: Design, manufacturing, simulation and mechanical testing. *Mater. Des.* **2017**, *114*, 546–556. [CrossRef]

13. Kumar, A.; Nune, K.C.; Murr, L.E.; Misra, R.D.K. Biocompatibility and mechanical behaviour of three-dimensional scaffolds for biomedical devices: Process-structure-property paradigm. *Int. Mater. Rev.* **2016**, *61*, 20–45. [CrossRef]

14. Harrison, N.; McHugh, P.E.; Curtin, W.; Mc Donnell, P. Micromotion and friction evaluation of a novel surface architecture for improved primary fixation of cementless orthopaedic implants. *J. Mech. Behav. Biomed. Mater.* **2013**, *21*, 37–46. [CrossRef] [PubMed]

15. Jetté, B.; Brailovski, V.; Dumas, M.; Simoneau, C.; Terriault, P. Femoral stem incorporating a diamond cubic lattice structure: Design, manufacture and testing. *J. Mech. Behav. Biomed. Mater.* **2018**, *77*, 58–72. [CrossRef] [PubMed]

16. Marin, E.; Fusi, S.; Pressacco, M.; Paussa, L.; Fedrizzi, L. Characterization of cellular solids in Ti6Al4V for orthopaedic implant applications: Trabecular titanium. *J. Mech. Behav. Biomed. Mater.* **2010**, *3*, 373–381. [CrossRef] [PubMed]

17. Le Guéhennec, L.; Soueidan, A.; Layrolle, P.; Amouriq, Y. Surface treatments of titanium dental implants for rapid osseointegration. *Dent. Mater.* **2007**, *23*, 844–854. [CrossRef] [PubMed]

18. Khanna, R.; Kokubo, T.; Matsushita, T.; Nomura, Y.; Nose, N.; Oomori, Y.; Yoshida, T.; Wakita, K.; Takadama, H. Novel artificial hip joint: A layer of alumina on Ti-6Al-4V alloy formed by micro-arc oxidation. *Mater. Sci. Eng. C* **2015**, *55*, 393–400. [CrossRef] [PubMed]

19. Ramsden, J.J.; Allen, D.M.; Stephenson, D.J.; Alcock, J.R.; Peggs, G.N.; Fuller, G.; Goch, G. The design and manufacture of biomedical surface. *CIRP Ann. Manuf. Technol.* **2007**, *56*, 687–711. [CrossRef]

20. Emmelmann, C.; Scheinemann, P.; Munsch, M.; Seyda, V. Laser additive manufacturing of modified implant surfaces with osseointegrative characteristics. *Phys. Procedia* **2011**, *12*, 375–384. [CrossRef]

21. Paris, M.; Götz, A.; Hettrich, I.; Bidan, C.M.; Dunlop, J.W.C.; Razi, H.; Zizak, I.; Hutmacher, D.W.; Fratzl, P.; Duda, G.N.; et al. Scaffold curvature-mediated novel biomineralization process originates a continuous soft tissue-to-bone interface. *Acta Biomater.* **2017**, *60*, 64–80. [CrossRef] [PubMed]

22. Wang, Z.; Wang, C.; Li, C.; Qin, Y.; Zhong, L.; Chen, B.; Li, Z.; Liu, H.; Chang, F.; Wang, J. Analysis of factors influencing bone ingrowth into three-dimensional printed porous metal scaffolds: A review. *J. Alloys Compd.* **2017**, *717*, 271–285. [CrossRef]

23. Schouman, T.; Schmitt, M.; Adam, C.; Dubois, G.; Rouch, P. Influence of the overall stiffness of a load-bearing porous titanium implant on bone ingrowth in critical-size mandibular bone defects in sheep. *J. Mech. Behav. Biomed. Mater.* **2016**, *59*, 484–496. [CrossRef] [PubMed]

24. de Wild, M.; Zimmermann, S.; Rüegg, J.; Schumacher, R.; Fleischmann, T.; Ghayor, C.; Weber, F.E. Influence of microarchitecture on osteoconduction and mechanics of porous titanium Scaffolds generated by selective laser melting. *3D Print. Addit. Manuf.* **2016**, *3*, 142–151. [CrossRef]

25. Taniguchi, N.; Fujibayashi, S.; Takemoto, M.; Sasaki, K.; Otsuki, B. Effect of pore size on bone ingrowth into porous titanium implants. *Mater. Sci. Eng. C* **2016**, *59*, 690–701. [CrossRef] [PubMed]

26. Jetté, B.; Brailovski, V.; Simoneau, C.; Dumas, M.; Terriault, P. Development and in vitro validation of a simplified numerical model for the design of a biomimetic femoral stem. *J. Mech. Behav. Biomed. Mater.* **2017**, *77*, 539–550. [CrossRef] [PubMed]

27. Bellini, C.M.; Galbusera, F.; Ceroni, R.G.; Raimondi, M.T. Loss in mechanical contact of cementless acetabular prostheses due to post-operative weight bearing: A biomechanical model. *Med. Eng. Phys.* **2007**, *29*, 175–181. [CrossRef] [PubMed]

28. Souffrant, R.; Zietz, C.; Fritsche, A.; Kluess, D.; Mittelmeier, W.; Bader, R. Advanced material modelling in numerical simulation of primary acetabular press-fit cup stability. *Comput. Methods Biomech. Biomed. Engin.* **2012**, *15*, 787–793. [CrossRef] [PubMed]

29. Small, S.R.; Berend, M.E.; Howard, L.A.; Rogge, R.D.; Buckley, C.A.; Ritter, M.A. High initial stability in porous titanium acetabular cups: A biomechanical study. *J. Arthroplast.* **2013**, *28*, 510–516. [CrossRef] [PubMed]

30. Udofia, I.; Liu, F.; Jin, Z.; Roberts, P.; Grigoris, P. The initial stability and contact mechanics of a press-fit resurfacing arthroplasty of the hip. *J. Bone Jt. Surg. Br.* **2007**, *89*, 549–556. [CrossRef] [PubMed]

31. Chang, J.-D.; Kim, T.-Y.; Rao, M.B.; Lee, S.-S.; Kim, I.-S. Revision total hip arthroplasty using a tapered, press-fit cementless revision stem in elderly patients. *J. Arthroplast.* **2011**, *26*, 1045–1049. [CrossRef] [PubMed]

32. Chanlalit, C.; Fitzsimmons, J.S.; Shukla, D.R.; An, K.-N.; O'Driscoll, S.W. Micromotion of plasma spray versus grit-blasted radial head prosthetic stem surfaces. *J. Shoulder Elb. Surg.* **2011**, *20*, 717–722. [CrossRef] [PubMed]

33. Le Cann, S.; Galland, A.; Rosa, B.; Le Corroller, T.; Pithioux, M.; Argenson, J.N.; Chabrand, P.; Parratte, S. Does surface roughness influence the primary stability of acetabular cups? A numerical and experimental biomechanical evaluation. *Med. Eng. Phys.* **2014**, *36*, 1185–1190. [CrossRef] [PubMed]

34. Goriainov, V.; Jones, A.; Briscoe, A.; New, A.; Dunlop, D. Do the cup surface properties influence the initial stability? *J. Arthroplast.* **2014**, *29*, 757–762. [CrossRef] [PubMed]

35. Gebert, A.; Peters, J.; Bishop, N.E.; Westphal, F.; Morlock, M.M. Influence of press-fit parameters on the primary stability of uncemented femoral resurfacing implants. *Med. Eng. Phys.* **2009**, *31*, 160–164. [CrossRef] [PubMed]

36. Ries, M.D.; Harbaugh, M.; Shea, J.; Lambert, R. Effect of cementless acetabular cup geometry on strain distribution and press-fit stability. *J. Arthroplast.* **1997**, *12*, 207–212. [CrossRef]

37. Adler, E.; Stuchin, S.A.; Kummer, F.J. Stability of press-fit acetabular cups. *J. Arthroplast.* **1992**, *7*, 295–301. [CrossRef]

38. Macdonald, W.; Carlsson, L.V.; Charnley, G.J.; Jacobsson, C.M. Press-fit acetabular cup fixation: Principles and testing. *Proc. Inst. Mech. Eng. Part H J. Eng. Med.* **1999**, *213*, 33–39. [CrossRef] [PubMed]

39. Morlock, M.; Götzen, N.; Sellenschloh, K. Bestimmung der Primärstabilität von künstlichen Hüftpfannen. In *DVM Bericht 314—Eigenschaften und Prüftechniken mechanisch Beanspruchter Implantate*; DVM: Berlin, Germany, 2002; pp. 221–229.

40. Toossi, N.; Adeli, B.; Timperley, A.J.; Haddad, F.S.; Maltenfort, M.; Parvizi, J. Acetabular components in total hip arthroplasty: Is there evidence that cementless fixation is better? *J. Bone Jt. Surg.* **2013**, *95*, 168–174. [CrossRef] [PubMed]

41. Roth, A.; Winzer, T.; Sander, K.; Anders, J.O.; Venbrocks, R.-A. Press fit fixation of cementless cups: How much stability do we need indeed? *Arch. Orthop. Trauma Surg.* **2006**, *126*, 77–81. [CrossRef] [PubMed]

42. Tabata, T.; Kaku, N.; Hara, K.; Tsumura, H. Initial stability of cementless acetabular cups: Press-fit and screw fixation interaction—An in vitro biomechanical study. *Eur. J. Orthop. Surg. Traumatol.* **2015**, *25*, 497–502. [CrossRef] [PubMed]

43. Takao, M.; Nakamura, N.; Ohzono, K.; Sakai, T.; Nishii, T.; Sugano, N. The results of a press-fit-only technique for acetabular fixation in hip dysplasia. *J. Arthroplast.* **2011**, *26*, 562–568. [CrossRef] [PubMed]

44. Amirouche, F.; Solitro, G.; Broviak, S.; Gonzalez, M.; Goldstein, W.; Barmada, R. Factors influencing initial cup stability in total hip arthroplasty. *Clin. Biomech.* **2014**, *29*, 1177–1185. [CrossRef] [PubMed]

45. Clarke, H.J.; Jinnah, R.H.; Warden, K.E.; Cox, Q.G.; Curtis, M.J. Evaluation of acetabular stability in uncemented prostheses. *J. Arthroplast.* **1991**, *6*, 335–340. [CrossRef]

46. Klanke, J.; Partenheimer, A.; Westermann, K. Biomechanical qualities of threaded acetabular cups. *Int. Orthop.* **2002**, *26*, 278–282. [CrossRef] [PubMed]

47. Baleani, M.; Fognani, R.; Toni, A. Initial stability of a cementless acetabular cup design: Experimental investigation on the effect of adding fins to the rim of the cup. *Artif. Organs.* **2001**, *25*, 664–669. [CrossRef] [PubMed]

48. Olory, B.; Havet, E.; Gabrion, A.; Vernois, J.; Mertl, P. Comparative in vitro assessment of the primay stbility of cementless press-fit acetabular cups. *Acta Orthop. Belg.* **2004**, *70*, 31–37. [PubMed]

49. Fritsche, A.; Zietz, C.; Teufel, S.; Kolp, W.; Tokar, I.; Mauch, C.; Mittelmeier, W.; Bader, R. In-vitro and in-vivo investigations of the impaction and pull-out behavior of metal-backed acetabular cups. *Br. Ed. Soc. Bone Jt. Surg.* **2011**, *93*, 406.

50. Weißmann, V.; Boss, C.; Bader, R.; Hansmann, H. A novel approach to determine primary stability of acetabular press-fit cups. *J. Mech. Behav. Biomed. Mater.* **2018**, *80*, 1–10. [CrossRef] [PubMed]

51. Markhoff, J.; Wieding, J.; Weissmann, V.; Pasold, J.A.; Jonitz-Heincke, R. Bader, Influence of different three-dimensional open porous titanium scaffold designs on human osteoblasts behavior in static and dynamic cell investigations. *Materials* **2015**, *8*, 5490–5507. [CrossRef] [PubMed]

52. Weißmann, V.; Bader, R.; Hansmann, H.; Laufer, N. Influence of the structural orientation on the mechanical properties of selective laser melted Ti6Al4V open-porous scaffolds. *Mater. Des.* **2016**, *95*, 188–197. [CrossRef]

53. Weißmann, V.; Wieding, J.; Hansmann, H.; Laufer, N.; Wolf, A.; Bader, R. Specific yielding of selective laser-melted Ti6Al4V open-porous scaffolds as a function of unit cell design and dimensions. *Metals* **2016**, *6*, 166. [CrossRef]

54. Weißmann, V.; Hansmann, H.; Bader, R.; Laufer, N. Influence of the Structural Orientation on the Mechanical Properties of Selective Laser Melted TiAL6V4 Open-Porous Scaffold. In Proceedings of the 13th Rapid Tech Conference Erfurt, Erfurt, Germany, 14–16 June 2016.

55. Fox, J.C.; Moylan, S.P.; Lane, B.M. Effect of process parameters on the surface roughness of overhanging structures in laser powder bed fusion additive manufacturing. *Procedia CIRP* **2016**, *45*, 131–134. [CrossRef]

56. Rashed, M.G.; Ashraf, M.; Mines, R.A.W.; Hazell, P.J. Metallic microlattice materials: A current state of the art on manufacturing, mechanical properties and applications. *Mater. Des.* **2016**, *95*, 518–533. [CrossRef]

57. Suard, M.; Martin, G.; Lhuissier, P.; Dendievel, R.; Vignat, F.; Blandin, J.J.; Villeneuve, F. Mechanical equivalent diameter of single struts for the stiffness prediction of lattice structures produced by Electron Beam Melting. *Addit. Manuf.* **2015**, *8*, 124–131. [CrossRef]

58. Weißmann, V.; Drescher, P.; Bader, R.; Seitz, H.; Hansmann, H.; Laufer, N. Comparison of single Ti6Al4V struts made using selective laser melting and electron beam melting subject to part orientation. *Metals* **2017**, *7*, 91. [CrossRef]

59. Triantaphyllou, A.; Giusca, C.L.; Macaulay, G.D.; Roerig, F.; Hoebel, M.; Leach, R.K.; Tomita, B.; Milne, K.A. Surface texture measurement for additive manufacturing. *Surf. Topogr. Metrol. Prop.* **2015**, *3*, 024002. [CrossRef]

60. Frosch, K.; Barvencik, F.; Viereck, V.; Lohmann, C.H.; Dresing, K.; Breme, J.; Brunner, E.; Stürmer, K.M. Growth behavior, matrix production, and gene expression of human osteoblasts in defined cylindrical titanium channels. *J. Biomed. Mater. Res. Part A* **2004**, *68*, 325–334. [CrossRef] [PubMed]

61. Knychala, J.; Bouropoulos, N.; Catt, C.J.; Katsamenis, O.L.; Please, C.P.; Sengers, B.G. Pore geometry regulates early stage human bone marrow cell tissue formation and organization. *Ann. Biomed. Eng.* **2013**, *41*, 917–930. [CrossRef] [PubMed]

62. Kienapfel, H.; Sprey, C.; Wilke, A.; Griss, P. Implant fixation by bone ingrowth. *J. Arthroplast.* **1999**, *14*, 355–368. [CrossRef]

63. Kawai, T.; Takemoto, M.; Fujibayashi, S.; Tanaka, M.; Akiyama, H.; Nakamura, T.; Matsuda, S. Comparison between alkali heat treatment and sprayed hydroxyapatite coating on thermally-sprayed rough Ti surface in rabbit model: Effects on bone-bonding ability and osteoconductivity. *J. Biomed. Mater. Res. Part B Appl. Biomater.* **2015**, *103*, 1069–1081. [CrossRef] [PubMed]

64. Grimal, Q.; Haupert, S.; Mitton, D.; Vastel, L.; Laugier, P. Assessment of cortical bone elasticity and strength: Mechanical testing and ultrasound provide complementary data. *Med. Eng. Phys.* **2009**, *31*, 1140–1147. [CrossRef] [PubMed]

65. Niinomi, M.; Nakai, M. Titanium-based biomaterials for preventing stress shielding between implant devices and bone. *Int. J. Biomater.* **2011**, *2011*. [CrossRef] [PubMed]

66. Wauthle, R.; Vrancken, B.; Beynaerts, B.; Jorissen, K.; Schrooten, J.; Kruth, J.-P.; Humbeeck, J. Effects of build orientation and heat treatment on the microstructure and mechanical properties of selective laser melted Ti6Al4 V lattice structures. *Addit. Manuf.* **2014**, *5*, 6–13. [CrossRef]

67. Goldman, A.H.; Armstrong, L.C.; Owen, J.R.; Wayne, J.S.; Jiranek, W.A. Does increased coefficient of friction of highly porous metal increase initial stability at the acetabular interface? *J. Arthroplast.* **2016**, *31*, 721–726. [CrossRef] [PubMed]

68. Ahmadi, S.M.; Campoli, G.; Amin Yavari, S.; Sajadi, B.; Wauthle, R.; Schrooten, J.; Weinans, H.; Zadpoor, A.A. Mechanical behavior of regular open-cell porous biomaterials made of diamond lattice unit cells. *J. Mech. Behav. Biomed. Mater.* **2014**, *34*, 106–115. [CrossRef] [PubMed]

69. Lopez-Heredia, M.A.; Goyenvalle, E.; Aguado, E.; Pilet, P.; Leroux, C.; Dorget, M.; Weiss, P.; Layrolle, P. Bone growth in rapid prototyped porous titanium implants. *J. Biomed. Mater. Res. Part A* **2008**, *85*, 664–673. [CrossRef] [PubMed]

70. Hedayati, R.; Sadighi, M.; Mohammadi-Aghdam, M.; Zadpoor, A.A. Mechanics of additively manufactured porous biomaterials based on the rhombicuboctahedron unit cell. *J. Mech. Behav. Biomed. Mater.* **2016**, *53*, 272–294. [CrossRef] [PubMed]

71. Gollwitzer, R.; Gradinger, H. *Ossäre Integration*; Springer Medizin Verlag: Heidelberg, Germany, 2006.

72. Swarts, E.; Bucher, T.A.; Phillips, M.; Yap, F.H.X. Does the ingrowth surface make a difference? A retrieval study of 423 cementless acetabular components. *J. Arthroplast.* **2015**, *30*, 706–712. [CrossRef] [PubMed]

Article

Digital Design, Analysis and 3D Printing of Prosthesis Scaffolds for Mandibular Reconstruction

Khaja Moiduddin *, Syed Hammad Mian, Hisham Alkhalefah and Usama Umer

Advanced Manufacturing Institute, King Saud University, Riyadh 11421, Saudi Arabia;
syedhammad68@yahoo.co.in (S.H.M.); halkhalefah@ksu.edu.sa (H.A.); usamaumer@yahoo.com (U.U.)
* Correspondence: kmoiduddin@gmail.com; Tel.: +96-611-469-7372

Received: 21 March 2019; Accepted: 14 May 2019; Published: 16 May 2019

Abstract: Segmental mandibular reconstruction has been a challenge for medical practitioners, despite significant advances in medical technology. There is a recent trend in relation to customized implants, made up of porous structures. These lightweight prosthesis scaffolds present a new direction in the evolution of mandibular restoration. Indeed, the design and properties of porous implants for mandibular reconstruction should be able to recover the anatomy and contour of the missing region as well as restore the functions, including mastication, swallowing, etc. In this work, two different designs for customized prosthesis scaffold have been assessed for mandibular continuity. These designs have been evaluated for functional and aesthetic aspects along with effective osseointegration. The two designs classified as top and bottom porous plate and inner porous plate were designed and realized through the integration of imaging technology (computer tomography), processing software and additive manufacturing (Electron Beam Melting). In addition, the proposed designs for prosthesis scaffolds were analyzed for their biomechanical properties, structural integrity, fitting accuracy and heaviness. The simulation of biomechanical activity revealed that the scaffold with top and bottom porous plate design inherited lower Von Mises stress (214.77 MPa) as compared to scaffold design with inner porous plate design (360.22 MPa). Moreover, the top and bottom porous plate design resulted in a better fit with an average deviation of 0.8274 mm and its structure was more efficiently interconnected through the network of channels without any cracks or powder material. Verily, this study has demonstrated the feasibility and effectiveness of the customized porous titanium implants in mandibular reconstruction. Notice that the design and formation of the porous implant play a crucial role in restoring the desired mandibular performance.

Keywords: mandibular reconstruction; scaffolds; reconstruction plate; finite element analysis; 3D printing; titanium alloy

---

1. Introduction

Mandibular reconstruction is recognized as the most challenging and significant procedures by maxillofacial surgeons. It can be attributed to the strict requirements demanded by patients, in terms of anatomy, outer profile of the mandible and optimal restoration of oral functions [1–4]. The problem of mandibular reconstruction is further escalated owing to a rapid increase in mandibular defects due to modern human skeletal diversity and chewing behavior [5]. Generally, the mandibular continuity defect involves a complete bone loss and is caused by infection, trauma, lesion, osteonecrosis and resection of benign and malignant tumors [1]. The timely and adequate rehabilitation of mandibular defect is crucial to prevent impairment of masticatory function, loss of speech, cosmetic deformity and to essentially maintain the patient's quality of life. Certainly, the titanium plate with autogenous bone transplantation can be regarded as the primary standard and a reliable treatment for mandibular reconstruction [6]. In spite of the availability of reconstruction techniques related to autogenous bone graft, perfect mandibular reconstruction is still not possible and remains a challenge. Generally, the available standard commercial reconstruction plates (implants) are employed in mandibular reformation. These plates are manufactured using traditional methods such as casting and the powder metallurgical process, which are time consuming processes [7]. Furthermore, the standard plates are straight and they need bending in order to align them along the mandible curved bone. This not only raises the operative (or surgery) time, but also involves the tedious task of repeatedly adapting and revising the plate according to the patient's anatomy. Since, it is a trial and error procedure, the possibility of discrepancies between the bone and plate interface increases, which in turn causes implant failure as well as discomfort to the patient. Therefore, it is indispensable to utilize custom made implants, which not only reduce disproportion and mismatch, but also result in improved appearance and actualization. The personalized implant design not only enhances fitting accuracy, but also minimizes the surgical time in contrast to standard plates.

Recent developments in tissue and scaffold engineering represents a contemporary prospect and a new application in the evolution of mandibular restoration. Scaffolds can be combined with solid parts and fabricated as an implant. Ideally, the scaffolds should be highly porous, crack free and biocompatible with tissue ingrowth [8]. As reported by numerous clinical studies, the titanium scaffold (porous structure) can achieve long term bone fixation and promote full bone ingrowth when compared to the solid or bulk part [9,10]. In addition, solid titanium implants due to variation in mechanical properties as compared to bone may lead to bone resorption, which induces stress shielding effect on its surrounding bone and eventually leads to implant failure [11]. The impeccable porosity influences cell behavior and the interconnected channels of pores stimulate the vascularization [12]. The encouragement of early osseointegration is critically important for the success of implantation, otherwise longer healing time would lead to implant failure [13].

With advancements in engineering technology, including medical modeling software and three-dimensional (3D) printing or additive manufacturing, it is now possible to design and fabricate customized implants with better accuracy and in a shorter period of time. The unification of data acquisition, image processing, as well as modeling and additive manufacturing, have made it possible to comprehend tailor-made implants according to the patient's requirements. Undoubtedly, the implementation of integrated techniques can save a lot of money for medical practitioners as well as revamp the quality of life for a large number of people [14]. The agreeable effect in mandible restoration depends on many aspects of the implant, including its design, fabrication technology, biomechanical properties, accuracy, surface integrity and weight. Certainly, 3D printing techniques have emerged as a promising potential in the development of bone reconstruction, rehabilitation and in the field of surgery [15]. Among several 3D printing techniques, electron beam melting (EBM) has been regarded as the fast and successful method for the fabrication of titanium medical implants from computer-aided design (CAD) models with Food and Drug Administration (FDA) and Conformité Européene (CE) approval [16]. EBM technique, which was first commercialized in 1997 by ARCAM AB, fabricate parts by melting metal powder in a layer-by-layer fashion [17]. It has increasingly been used for the fabrication of 3D titanium alloy scaffolds for medical applications with complex architecture [18,19]. Mandibular bone is not a uniform and regular structure, but rather a curved and special structure. Therefore, very few researchers have attempted to custom design prosthesis for mandibular reconstruction [20,21] and very limited information is available on the study of mandibular scaffold. In addition, no clear evidence and investigation are available in the biomechanical, structural integrity and fitting evaluation of mandibular prosthetic scaffolds.

In this study, two different types of custom specific mandibular prosthesis scaffolds have been designed, fabricated and evaluated for their performance. These two designs were categorized as top and bottom porous plate and inner porous plate. In the top and bottom porous plate design, the mesh or porous structure was attached on the top and bottom of the plate, whereas in the inner porous plate design, the porous structure was inside the plate. An extensive integrated methodology has been utilized for the realization of the patient-specific porous implant. The part fabrication using EBM was supplemented with computer tomography (CT) for image acquisition and processing software for implant modeling. The two scaffold designs were also analyzed to determine their biomechanical effect under the mastication process using Finite Element Analysis (FEA), surface integrity using micro-CT scans as well as fitting accuracy and appearance utilizing the 3D comparison technique.

2. Methodology

The typical flowchart as shown in Figure 1, demonstrates the methodology adopted in this work. It was based on six primary steps: Data acquisition, customized implant design and modeling, virtual assembly, FEA, part fabrication and evaluation. This approach was prominent because it involved interaction between the engineering and medical fields right from the patient diagnosis until the mandibular reconstruction. The authors in this methodology have emphasized the importance of communication between the engineering and medical departments. In the current study, the medical practitioners were customers, therefore, they were engaged in each and every stage during the entire process. These communication links are evidently specified by using red circles in the Figure 1. These communications acted as a feedback loop to get the assessment or the criticism from the medical people. Of course, the engineers had to explain various aspects and engineering terms or analysis to medical professionals before every session. This communication or information exchange helped to improve the overall results by minimizing design revision and preventing implant failure.

2.1. Data Acquisition

A forty-year-old patient with deformities and a lesion in the left mandibular area attended the emergency department of the university hospital. Upon diagnosis and a series of tests by the medical doctor, the patient was subjected to a non-invasive CT scans. The non-invasive CT can be defined as a medical procedure which does not involve any deterioration of the skin, internal body as well as the destruction of healthy tissues. During the course of patient diagnosis, it was found that the patient was suffering from mandibular continuity defect with a loss of portion of the bone resulting in a gap of ~2 cm or more. It is a patient-specific defect which is larger in size. The CT images were acquired using a Promax 3D "Cone beam computer tomography machine" (Planmeca, Helsinki, Finland) [22]. The minimum resolution model (voxel size) was $0.10~mm^3$. It was implemented under the following conditions: Voltage—54–90 kV, Current—1–14 mA, Focal spot 0.4 mm, detector resolution 127 μm, scan time 18–26 s. The radiologist performed the CT scan on the patient and saved the scanned images in Digital Imaging and Communications in Medicine (DICOM) format which is a universal stored format for medical images. The DICOM files containing a series of two-dimensional (2D) images, stored in a database, did not provide a perfect picture of the anatomical structure. Several medical modeling and image processing software available in the market were used to convert the 2D images into a 3D anatomical model. MIMICS 17.0® (Materialise Interactive Medical Image Control System; Materialise NV, Leuven, Belgium) was used in this study. The 2D images of DICOM files were imported into MIMICS® which stacked the 2D images over each other and developed a typical 3D model. In medical CT imaging, the Hounsfield unit (HU) represents the grayscale from black to white with a range from −1024 (minimum value) to 3071 (maximum value). A custom thresholding Hounsfield unit of 282 to 2890 HU was used for bone identification. Segmentation by thresholding technique was used to select the soft and hard tissue by defining the range of the threshold value. Figure 2 illustrates the patient mandibular tumor in a different view.

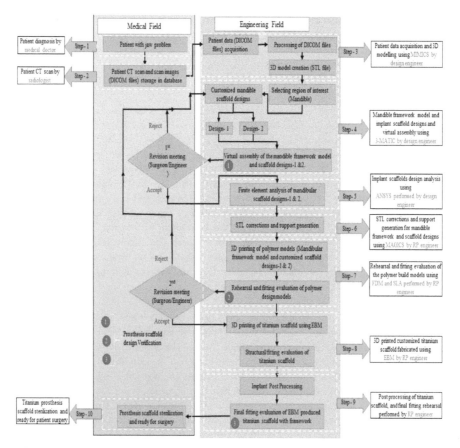

Figure 1. The proposed methodology for design, analysis and fabrication of customized mandibular prosthesis scaffolds. Note: The red circles indicate the formal meetings between the engineering and medical department for scaffold design verification and evaluation.

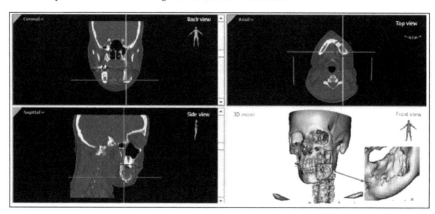

Figure 2. Patient anatomical model depicting the tumor region in different planes.

2.2. Customized Implant Design and Modeling

The region growing technique using MIMICS was used to extract the region of interest (mandible) from the surrounding tissues. Figure 3a–e illustrates the region growing techniques, where the full face mask was segregated to the region of interest in mandible Figure 3e. The obtained tumor mandible without teeth was then saved as a Standard Tessellation Language (STL) file. The STL file was imported into 3-Matic® (Materialise, Leuven, Belgium) for implant design. Mirror reconstruction design technique is the most common implant design where the healthy bone is mirrored and replaced over the defective bone. Several research studies have proved that mirror reconstruction technique has successfully restored and provided excellent facial symmetry [23,24]. The tumor on the left mandible (Figure 3f) was resected and the right side of the healthy mandibular bone was mirrored as shown in Figure 3g. The symmetrical sides were merged to form a healthy mandible. Wrapping operation was performed to nullify the gaps and voids. The obtained healthy mandible (Figure 3h) was used for the implant design by selecting (Figure 3i) and extracting the outer region (Figure 3j) for customized implant design. Smoothing and trimming operations were performed to get the implant design shape as shown in Figure 3k. An offset thickness of 2 mm (Figure 3l) was provided and two implant designs with one inner bone graft carrier and the other with top and bottom bone graft carrier were designed as shown in Figure 3m,m′. The inner plate and thick top and bottom plate were patterned into the porous structure (scaffold) using dode thick (Figure 3n) from Magics® (Materialise, Belgium) as shown in Figure 3o. The dode thick mesh structure was used to reduce the weight of the mandibular implant and to provide good adhesion between the bone and the implant. Several research articles have proved that titanium scaffold with a porosity of 500–1000 microns influence the osseointegration and faster bone healing [25,26]. Figure 3p illustrates the designed scaffold pore (900 microns) and strut (300 microns) size.

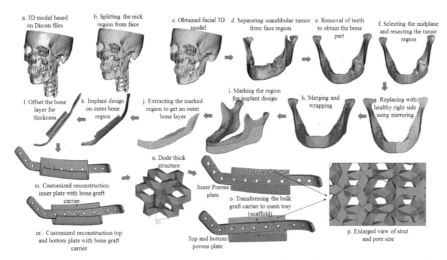

Figure 3. Sequence of steps in the design of customized prosthesis scaffold (implant) for mandibular defects.

2.3. Virtual Assembly

The two designed prosthesis scaffolds were virtually assembled and aligned with the mandibular framework model for fitting and assembly evaluation as shown in Figure 4. Formal meetings used to take place between the engineering and medical field for evaluating and verifying the design as indicated by red circles (Figure 1). Any error or void in-between the implant and the bone would result in the redesigning of the implant. The virtual assembly also helped with surgical guidance, understanding the surgical anatomy and real world preoperative surgery scenario to improve the reliability and safety of the surgical process.

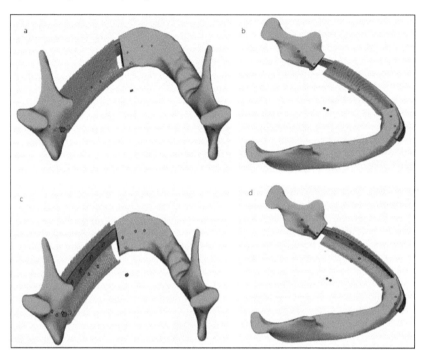

Figure 4. Posterior (back) and top view of the two customized scaffolds: Inner porous plate (**a,b**) and (**c,d**) top and bottom porous plate.

The designed reconstruction scaffolds were incorporated with countersink medical screw holes with three screws on the condyle side and three screws on the chin area. The countersink holes were designed for the complete immersion of the screw head inside the screw hole in order to provide a better aesthetic effect. Figure 5 illustrates the virtual assembly of the mandibular framework model containing the cortical and trabecular bone with scaffold fitted with six screws. The error free designed scaffold and the framework model were saved as a Standard for the Exchange of Product model data (STP) file for analysis.

2.4. Finite Element Analysis

Once the designed scaffolds were examined for fitting and conformance in the virtual assembly, the FEA model was created to evaluate their functionality as well as the biomechanical effect of clenching on the prosthesis scaffold. The FEA was employed because it is recognized as one of the crucial tools to emulate and predict the behavior of the CAD model in real scenarios. It was first used in the aerospace industry but quickly spread throughout a wide range of sciences including medicine and dentistry [27]. A finite element model (FEM) consisting of the temporomandibular model and two designed scaffolds was created using Ansys® software. In this study, the sustained clenching and masticatory muscle activity using three muscular forces (masseter, medial pterygoid and temporalis) were simulated. The material properties of the cortical bone, trabecular bone, screws and scaffold were adapted from the literature study and were assumed as homogeneous, isotropic and linear elastic [28,29]. The Young's modulus, Poisson's ratio and yield strength of the simulated study are presented in Table 1.

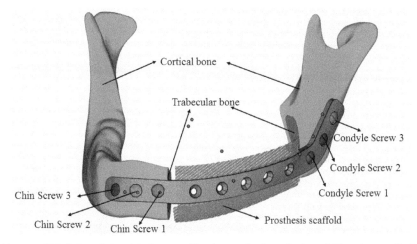

Figure 5. Global view of virtual design assembly of customized prosthesis scaffold on the mandibular framework model.

Table 1. Mechanical properties of study materials used in FE model. Data from [28,29].

Materials	Young's Modulus (MPa)	Poisson's Ratio	Yield Strength (MPa)
Compact Bone	13,700	0.3	122
Trabecular Bone	1370	0.3	2
Prosthesis scaffold, (Ti6Al4V ELI)	120,000	0.3	930

For clenching simulation, the superior part of both condyles was constrained in all directions. The displacement in the molar region as shown in Figure 6 was restrained in the upper region to simulate chewing. While the biting forces acted axially, the molar movement was kept at near zero displacement. This restraint was perpendicular to the occlusal plane (Z-direction), while allowing freedom of movement in the horizontal plane (X and Y direction). The FEM was meshed with the 10-node 3D tetrahedral element.

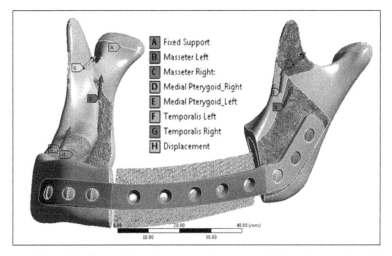

Figure 6. Typical loading and boundary constraints on mandibular framework model with prosthesis scaffold.

As shown in Figure 7, the triangle surface mesher strategy with program controlled patch conforming method was used in order to refine the mesh at the area of fixation and to obtain more accurate results. The magnitude and boundary condition of the masticatory forces were derived from the literature study [30,31]. The interface between the scaffold-bone and screw-scaffold-bone were considered as bonded. The clenching movement was simulated in the FEM with muscular forces and their vectors are presented in Table 2.

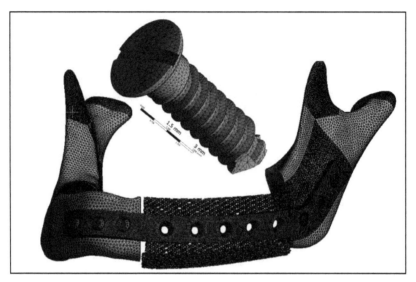

Figure 7. Meshing on the simulated mandibular framework model with prosthesis scaffold and a close-up view of screw meshing.

Table 2. Magnitude and functional direction of masticatory muscles in Newton's (N). Data from [30,31].

Masticatory Muscles	X (N)	Y (N)	Z (N)
Masseter	50	−50	200
Medial pterygoid	0	−50	100
Temporalis	0	100	200

2.5. Fabrication

In this study, 3D printing was used for the fabrication of customized prosthesis scaffolds. Two types of materials—polymer and metal—were used in the fabrication. The polymer 3D printing was used for the testing and fitting evaluation (virtual assembly), whereas metal (Ti6Al4V ELI) was used for the patient prosthesis implant. For polymer-based 3D printing, Stratasys-fused deposition modeling (FDM) machine and FORMLABS-2 a (stereolithography) SLA machine were used. ARCAM's EBM machine (EBM A2, ARCAM AB, Mölndal, Sweden) was used for printing titanium metal scaffolds.

2.5.1. Polymer Fabrication

The FDM machine as shown in Figure 8a was used to print mandibular framework models (Figure 8b) using ABS (acrylonitrile butadiene styrene) material which is a common thermoplastic resin with good functional properties [32]. FDM works on additive manufacturing process where the ABS material unwound from the coil and is heated to melting point and extruded in a layer-by-layer fashion to produce 3D objects. Formlabs-2 3D printer as shown in Figure 8c was used to fabricate the mandibular prosthesis scaffold (Figure 8d) which used the liquid resin material. Formlabs-2 form works on laser-based SLA principle where the laser solidifies the liquid resin material in a photo-polymerization process and builds the 3D model in a layer-by-layer fashion [33]. SLA produces objects with higher resolution with more accuracy when compared to FDM due to its optimal spot size laser which is very small [34]. Formlabs-2 was used in the fabrication of mandibular scaffold as it provided higher resolution and accuracy for the complicated porous structures.

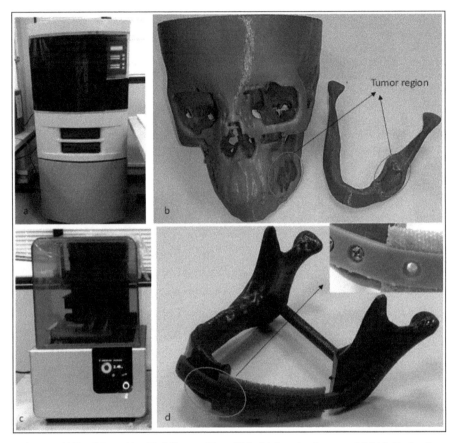

Figure 8. (**a**) Fused Deposition Modeling machine with its fabricated polymer model (**b**) indicating the tumor region and (**c**) SLA machine and its produced mandibular scaffold (**d**) with a close-up view.

2.5.2. Titanium Fabrication

It is well proven that scaffolds with elastic modulus closer to that of bone, minimizes the stress shielding effect and promotes bone-implant tissue in-growth [35,36]. Powder bed metal based 3D printing technologies such as EBM and selective laser melting (SLM) have demonstrated the capability to produce scaffolds in medical applications [37]. The EBM process in comparison requires less supporting material and minimizes post processing steps such as machining and heat treatment [36]. An EBM process is most suited for reactive metals such as titanium alloy as the complete build process takes place in a vacuum environment [38]. In addition, EBM produces parts at a much faster rate (80 cm^3/h) when compared to SLM (20–40 cm^3/h) [39]. The standard layer thickness of the printed samples using ARCAM's A2 EBM machine was 50–70 μm.

Figure 9a,b illustrates the typical working principle of the EBM process and the different components of the EBM machine respectively. The tungsten filament in the electron beam gun on reaching above 2500 °C, emits a beam of electrons which accelerates at half the speed of light and passes through a series of controlled coils (lens) and impacts the powder surface, thus melting the powder. The first (astigmatism) lens assists to keep the beam in circular and round shape regardless of its position on the build plate. Without this coil, the focus point of the beam tends to have a wider area (elliptical shape) when it is deflected towards the edge of the build region. It also eliminates electro-optical artifacts (human error). The second (focus) lens keeps the beam in focus and sharpens to a desired (0.1 mm) diameter. The third (deflection) lens scans the beam across the build area. The build process takes place inside the build chamber. Inside the build chamber, there are two hoppers which hold the metal stock powder. Metal powder is spread homogeneously over the build table using rakes. The rakes fetches the powder from either end of hoppers and spreads it evenly over the build table. The build tank lowers down in the z-direction after each melt cycle. The start plate was placed at the center of the build table which holds the build surrounded by powder. Vacuum is maintained throughout the build cycle to eliminate impurities and to prevent reactions between the reactive metals. Titanium powder (Ti6Al4V ELI) with the particle size of 50–100 mm was used in this study. The chemical composition of Ti6Al4V ELI (extra low interstitial) was made of 6.04% Al, 4.05% V, 0.013% C, 0.0107% Fe, and 0.13% O, while the rest as Titanium (in weight percent).

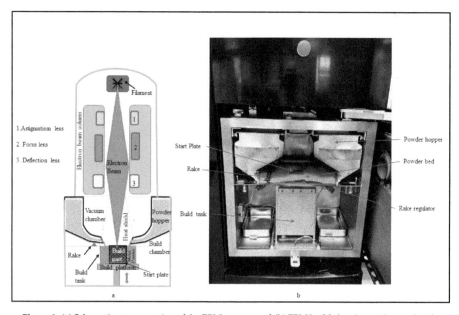

Figure 9. (**a**) Schematic representation of the EBM process and (**b**) EBM build chamber with part details.

The part fabrication in the EBM machine (ARCAM A2) as shown in Figure 10b is dependent on three phases—(1) Preheating of the metal powder. (2) Scanning and melting. (3) Lowering of build table and raking of powder.

(1). Preheating the metal powder: The Ti6Al4V ELI metal powder spread on the powder bed is preheated by multiple beams of electron at high scan speed and low beam current to reduce the internal residual stresses.

(2). Scanning and melting: The high velocity beam of electrons scans the metal powder and melts the power in line as per the defined CAD geometry. The melting process consist of two steps, melting the contours (outer and inner boundary) and infill hatching. The majority of the melting takes place in hatching where the beam current and scan speed are increased.

(3). Lowering build table and raking of powder: The build table is lowered after each melt layer cycle (50 µm) and a new layer of powder is fed from hoppers and spread evenly on the previously solidified powder layer using rakes. This process continues till the final 3D part is built.

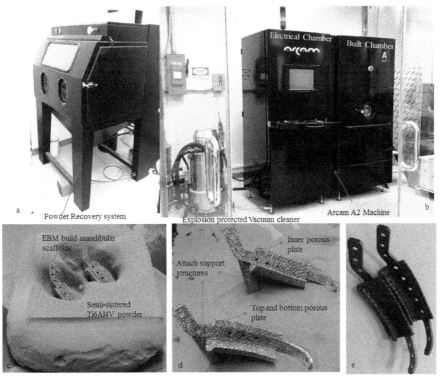

Figure 10. (**a**) PRS machine, (**b**) EBM machine with explosion protection vacuum cleaner, (**c**) EBM built mandibular prosthesis scaffold surrounded by semi-sintered powder, (**d**) titanium scaffolds with support structures and (**e**) mandibular scaffolds after support removal.

The EBM build lasted approximately 8–10 h. After build completion, the produced part (mandibular prosthesis scaffold) was allowed to cool under helium gas. Figure 10c shows the EBM build scaffold with supports surrounded by semi-sintered powder. The semi-sintered titanium powder was then blasted in powder recovery system (PRS) as shown in Figure 10a as a post processing process and to get the finished part with supports. The supports (Figure 10d) which were added to the scaffolds during the build to dissipate the heat and the overhang structures were manually removed with simple tools such as pliers. Figure 10e illustrates the final EBM built mandibular scaffolds which can be sandblasted or machined using laser ablation to achieve a smoother finish if required [40].

2.6. Evaluation and Validation

At this stage, the fabricated titanium scaffolds were investigated for structural integrity, fitting accuracy as well as the weight.

2.6.1. Micro-CT Scan on Titanium Lattice Structure

A non-destructive technique (i.e., micro-CT scan) was employed in order to examine the stochastic defects and structural integrity of the dode thick mesh structure used in scaffold design. The micro-CT scans were utilized in order to validate the quality of the dode thick structure in terms of cracks, internal trapped powder, in addition to examine the interior construction of the built struts without any physical cutting and polishing. A 15 mm solid cube (Figure 11a) was designed and transformed into a dode thick structure (Figure 11b,c) and fabricated using EBM as shown in Figure 11d. The micro-CT scanner (Bruker Skycam 1173, Kontich, Belgium) with a source voltage of 120 KV focused on the EBM fabricated cube structure with a spot size of 5 μm and with an image pixel size of 12.03 μm. Each 2D slice image of the cubic structure in the form of 512 × 512 bitmaps as output data was collected.

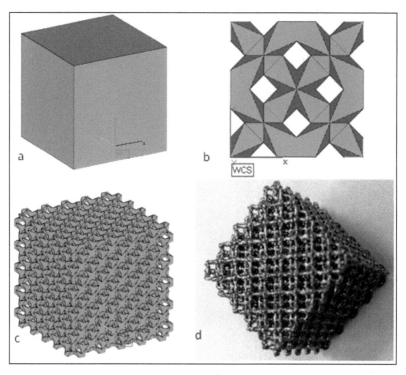

Figure 11. Cubes with unit cell structure of 15 × 15 mm^2 (a) solid cube, (b) dode thick unit cell structure, (c) dode thick cube structure and (d) EBM fabricated dode thick cube.

2.6.2. 3D Comparison

The 3D comparison technique was implemented in order to accurately compare the fitting accuracy of both the implant designs (inner porous plate and top and bottom porous plate) with respect to the mandible. The fitting accuracy of the implants was computed using Geomagics Control® [41]. The 3D comparison analysis can be considered as one of the most powerful and extensive techniques, to graphically represent the surface deviations between the reconstructed objects and the reference CAD model [42]. At the outset, the test model had to be aligned on the reference CAD model by utilizing the best fit alignment. Consequently, the analysis software automatically estimated the best fit between the test and reference object. This best fit alignment confirmed that both the test and reference objects were positioned (or fixed) in the same coordinate system. Furthermore, the statistic used in this work in order to quantify the fitting accuracy of the implants on the mandible was the average deviation. This statistic was utilized because it reported the deviation in the mandible, thereby approximating the gap between the implant (scaffold) and the mandible. In this work, the test model was acquired as a point cloud set by employing the laser scanner mounted on the Faro Platinum arm (FARO, Lake Mary, FL, USA) as shown in Figure 12.

Figure 12. Acquisition of test data using a Faro Platinum arm.

As shown in Figure 13, the scaffolds were mounted on the mandible and scanned to obtain the test data. The reference model was obtained by removing the defect and imitating the healthy side on it. The reference model acquired using the mirroring technique was assumed to represent the ideal anatomical structure [23,24].

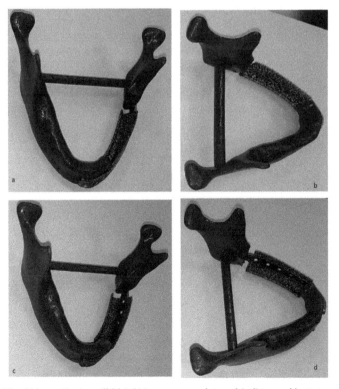

Figure 13. Mandible prosthesis scaffold (**a**,**b**) inner porous plate and (**c**,**d**) top and bottom porous plate mounted on the mandibular framework.

The outer surface of the scaffold mounted mandible were scanned and imported as STL model in Geomagics control® in order to compare it with the reference mandible. The outer surface was studied because the customized scaffolds were designed depending on the outer profile of the mandible. The 3D comparison analysis software represented the result by means of error scale through the computation of the shortest distance between the test model and the surface of the reference model.

2.6.3. Weights of the Scaffold Designs

In order to reduce the stress shielding effect between the implant and the surrounding bone, it was imperative to build lighter implants with weights closer to that of the bone being replaced [43]. The minimization of stress shielding was critical for reducing bone resorption as well as decreasing the rate of aseptic loosening. The weight of the mandibular bone to be replaced was calculated from the density formulae where volume was taken from the Magics® software (Materialise, Leuven, Belgium) and assuming density as 1600 kg/m³ [44]. The weights of the two EBM fabricated scaffolds were measured using a digital weighing machine.

3. Results and Discussion

In this work, two customized prosthesis scaffolds were designed from the patient CT scan files. The clinical setup for both the designed scaffolds were simulated under physiological clenching conditions. The FEA analysis was essential in order to find out the continuous grabbing and chewing ability of the designed customized implants. The equivalent stresses and strains observed on both scaffolds are presented in Figure 14. The results indicated that the maximum stresses in both customized scaffolds were confined to the mesh structure and it was evident due to its lower cross sectional area.

The simulated result summary of both designed scaffolds is presented in Table 3. The analysis showed that the FEA of inner porous plate design induced higher stress concentration than the FEA of top and bottom porous plate design. In addition, the maximum stresses on both the prosthesis scaffolds were well below the yield strength (930 MPa) of the titanium alloy (Ti6Al4V ELI). On further observation, the analysis results of the screws, revealed that the condyle screws exhibited higher stresses when compared to chin screws which indicated that the stresses were transferring from the bottom chin region towards the condyle side thus satisfying the mastication process [45].

Table 3. Summary of Von Mises stress, strain and deformation of two designed scaffolds.

				Stress on Chin Screw (MPa)			Stress on Condyl Screw (MPa)		
	FEA Outcomes					Screw Numbers			
Designed Implant	Max Von Mises Stress (MPa)	Max Strain	Deformation	1	2	3	1	2	3
Inner porous plate	360.22	0.0032	0.29852	55.85	38.26	50.52	122.9	121.74	81.5
Top & bottom porous plate	214.77	0.0068	0.31711	61.85	39.76	53.61	127.71	125.07	84.44

The most common cause for the failure of the mandibular reconstruction is either due to the reconstruction plate failure (excessive loads) or instability in the anchoring of the screws. In this study, the maximum stresses were found to be on the scaffold rather than on the screws and were well below the yield point and fatigue strength of the material. The stresses found on the screws in both the FEM were quiet less and within the failure limits, with the highest stress observed on the top and bottom screw plate. The other important parameter of the reconstruction plate design is its flexibility, to absorb the forces and chewing load conditions. The max strain on the inner porous plate was found to be 3.2 microns and the top and bottom porous plate was 6.8 microns. The maximum strain obtained on both the designed scaffolds was less and few microns. Based on the FEA results, it seems more reasonable to use prosthesis based on the top and bottom porous plate design for mandibular reconstruction, though both the plates were mechanically stable for fixation and could bear the masticatory functions.

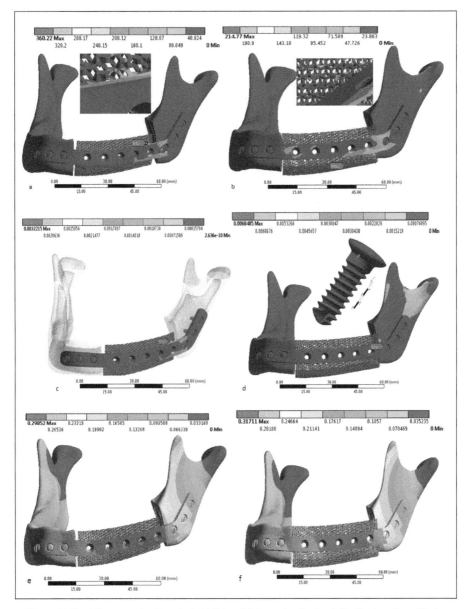

Figure 14. Von Mises stress (**top**), strain (**middle**) and deformation (**bottom**) distribution of mandibular framework model with two scaffolds (**a,c,e**) inner porous and (**b,d,f**) top and bottom porous plate.

The micro-CT scan results as shown in Figure 15 indicated that the dode thick structure was interconnected by a series of network channels and was free from any substantial internal defects such as cracks or voids. Similar results can be assumed and expected for the EBM fabricated mandibular prosthesis scaffold with dode thick structure.

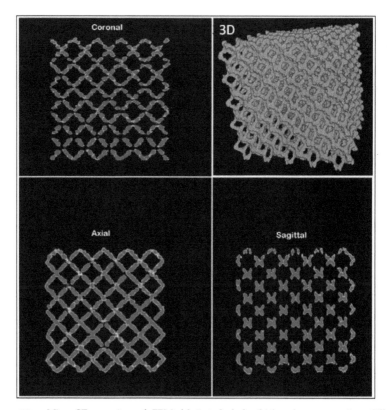

Figure 15. Micro-CT scanning of EBM fabricated dode thick cube representing different cross-sectional views.

The outcome of the 3D fitting deviation analysis has been represented graphically in Figure 16. The comprehensive investigation revealed that the scaffold with the top and bottom porous plate design provided better fitting accuracy as compared to the scaffold with inner porous plate design. An average deviation of 0.8274 mm was observed in the top and bottom porous plate design in comparison to 0.9283 mm of gap in the inner porous plate design.

The results of the weight analysis are presented in Table 4. The weight of the inner porous plate design was found to be 10.67 g and the top and bottom porous plate was 8.14 g. The weights of both reconstruction scaffolds were taken without considering the bone graft which will be placed inside the mesh carrier (tray) upon implant. Both scaffolds were low in weight and closer to that of bone properties. Certainly, this analysis confirmed that both the proposed designs possessed a lighter weight in comparison to their bone counterpart (19 g).

Table 4. Weight details of EBM fabricated scaffolds and replaced mandibular bone portion.

Parts	Replaced Bone	Inner Porous	Top and Bottom Porous
Volume (mm^3)	11879.00	2016.00	1847.00
Weight (g)	19.00	10.67	8.14

The Figure 17 illustrates the polymer and EBM fabricated titanium mandibular prosthesis scaffolds for final review before surgery.

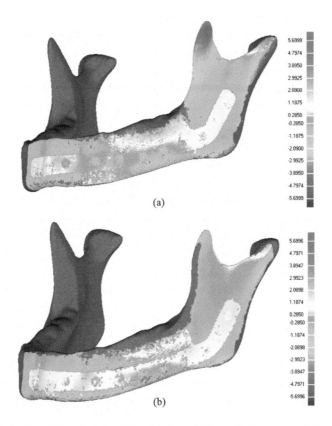

Figure 16. Evaluation of fitting deviation different designs: (**a**) Top and bottom porous; (**b**) inner porous.

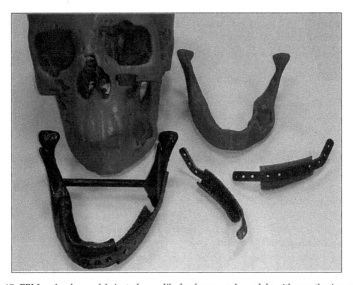

Figure 17. EBM and polymer fabricated mandibular framework models with prosthesis scaffolds.

4. Conclusions

The success of mandibular reconstruction greatly depends on its aesthetics and biomechanical properties. It emphasizes the importance of the customized implants depending on the patient's anatomy. The custom designed implants provide a better option for mandible restoration than the generic counterpart as they can fit precisely on the patient's bone. The ability to 3D print custom designed scaffolds using EBM technology, providing surface texture conducive to tissue ingrowth makes them appropriate for the personalized implants with properties closer to that of bone. In this study, two customized scaffolds based on the inner porous plate as well as the top and bottom porous plate were designed, 3D printed and evaluated for structural integrity, weight and fitting accuracy. A competent methodology has been presented to acquire the customized, pleasing and reliable mandibular implants. The methodology was exhaustive comprising of data acquisition using CT, mandible reconstruction as well as design, FEA, implant fabrication and testing.

Eventually, depending on the FEA, weight analysis and fitting accuracy evaluation, it can be inferred that the scaffold with the top and bottom porous plate is more favorable for bone reconstruction as compared to scaffold with the inner porous implant and can successfully be employed in the reconstruction of the defective mandible. Indeed, it can be asserted that the employment of prosthesis scaffolds in mandibular reconstruction satisfies the sustained need of lighter implants with accurate fitting and lesser surgical time and minimal revisions.

The customized porous implants are very effective and valuable because they provide an improved fit, enhanced osseointegration properties, lesser shielding effect and a higher implant stability. They strengthen the functional recovery of the mandibular deformities and maintain a graceful appearance on the mandible. It is mandatory that the research in this area should continue in the future for acquiring further innovative implant designs and reconstruction methods. The authors would like to expand this work by introducing new designs with different porous structures, and analyzing them for their strength and accuracy in mandible restoration. In addition, the authors would like to extend this work by including an extensive clinical (in-vivo) study in the future.

Author Contributions: K.M. conceived and designed the experiments; K.M. & S.H.M. performed the experiments; U.U. helped in the analysis; H.A. analyzed the data; K.M. & S.H.M. wrote and revised the paper.

Funding: This research was financially supported by Deanship of Scientific Research, King Saud University: Research group No. RG-1440-034.

Acknowledgments: The authors extend their appreciation to the Deanship of Scientific Research at King Saud University for funding this work through Research Group no. RG-1440-034.

Conflicts of Interest: The authors have no conflict of interest to declare.

References

1. Wong, R.C.W.; Tideman, H.; Kin, L.; Merkx, M.A.W. Biomechanics of mandibular reconstruction: a review. *Int. J. Oral Maxillofac. Surg.* **2010**, *39*, 313–319. [CrossRef]
2. Yan, R.; Luo, D.; Huang, H.; Li, R.; Yu, N.; Liu, C.; Hu, M.; Rong, Q. Electron beam melting in the fabrication of three-dimensional mesh titanium mandibular prosthesis scaffold. *Sci. Rep.* **2018**, *8*, 750. [CrossRef]
3. Miles, B.A.; Goldstein, D.P.; Gilbert, R.W.; Gullane, P.J. Mandible reconstruction. *Curr. Opin. Otolaryngol. Head Neck Surg.* **2010**, *18*, 317–322.
4. Hayden, R.E.; Mullin, D.P.; Patel, A.K. Reconstruction of the segmental mandibular defect: current state of the art. *Curr. Opin. Otolaryngol. Head Neck Surg.* **2012**, *20*, 231–236. [CrossRef]
5. Von Cramon-Taubadel, N. Global human mandibular variation reflects differences in agricultural and hunter-gatherer subsistence strategies. *Proc. Natl. Acad. Sci. USA* **2011**, *108*, 19546–19551. [CrossRef]
6. Yuan, J.; Cui, L.; Zhang, W.J.; Liu, W.; Cao, Y. Repair of canine mandibular bone defects with bone marrow stromal cells and porous β-tricalcium phosphate. *Biomaterials* **2007**, *28*, 1005–1013. [CrossRef] [PubMed]
7. Moiduddin, K. Implementation of Computer-Assisted Design, Analysis, and Additive Manufactured Customized Mandibular Implants. *J. Med. Biol. Eng.* **2018**, *38*, 744–756. [CrossRef]

8. Chanchareonsook, N.; Junker, R.; Jongpaiboonkit, L.; Jansen, J.A. Tissue-engineered mandibular bone reconstruction for continuity defects: A systematic approach to the literature. *Tissue Eng. Part B Rev.* **2014**, *20*, 147–162. [CrossRef] [PubMed]

9. Ryan, G.; Pandit, A.; Apatsidis, D. Fabrication methods of porous metals for use in orthopaedic applications. *Biomaterials* **2006**, *27*, 2651–2670. [CrossRef] [PubMed]

10. Wang, X.; Xu, S.; Zhou, S.; Xu, W.; Leary, M.; Choong, P.; Qian, M.; Brandt, M.; Xie, Y.M. Topological design and additive manufacturing of porous metals for bone scaffolds and orthopaedic implants: A review. *Biomaterials* **2016**, *83*, 127–141. [CrossRef] [PubMed]

11. Moiduddin, K.; Al-Ahmari, A.; Kindi, M.A.; Nasr, E.S.A.; Mohammad, A.; Ramalingam, S. Customized porous implants by additive manufacturing for zygomatic reconstruction. *Biocybern. Biomed. Eng.* **2016**, *36*, 719–730. [CrossRef]

12. Pei, X.; Zhang, B.; Fan, Y.; Zhu, X.; Sun, Y.; Wang, Q.; Zhang, X.; Zhou, C. Bionic mechanical design of titanium bone tissue implants and 3D printing manufacture. *Mater. Lett.* **2017**, *208*, 133–137. [CrossRef]

13. Raghavendra, S.; Wood, M.C.; Taylor, T.D. Early wound healing around endosseous implants: a review of the literature. *Int. J. Oral Maxillofac. Implant.* **2005**, *20*, 425–431.

14. Singare, S.; Lian, Q.; Wang, W.P.; Wang, J.; Liu, Y.; Li, D.; Lu, B. Rapid prototyping assisted surgery planning and custom implant design. *Rapid Prototyp. J.* **2009**, *15*, 19–23. [CrossRef]

15. Emadabouel, N.; Abdulrahman, A.-A.; Khaja, M.; Al Kindi, M.; Kamrani, A. A digital design methodology for surgical planning and fabrication of customized mandible implants. *Rapid Prototyp. J.* **2016**, *23*, 101–109.

16. Chua, C.K.; Wong, C.H.; Yeong, W.Y. *Standards, Quality Control, and Measurement Sciences in 3D Printing and Additive Manufacturing*; Academic Press: London, UK, 2017.

17. Electron Beam Melting—EBM Process, Additive Manufacturing. Available online: http://www.arcam.com/technology/electron-beam-melting/ (accessed on 7 July 2017).

18. Moiduddin, K.; Darwish, S.; Al-Ahmari, A.; El Watidy, S.; Mohammad, A.; Ameen, W. Structural and mechanical characterization of custom design cranial implant created using additive manufacturing. *Electron. J. Biotechnol.* **2017**, *29*, 22–31. [CrossRef]

19. Moiduddin, K.; Anwar, S.; Ahmed, N.; Ashfaq, M.; Al-Ahmari, A. Computer assisted design and analysis of customized porous plate for mandibular reconstruction. *IRBM* **2017**, *38*, 78–89. [CrossRef]

20. Narra, N.; Valášek, J.; Hannula, M.; Marcián, P.; Sándor, G.K.; Hyttinen, J.; Wolff, J. Finite element analysis of customized reconstruction plates for mandibular continuity defect therapy. *J. Biomech.* **2014**, *47*, 264–268. [CrossRef]

21. Liu, Y.; Fan, Y.; Jiang, X.; Baur, D.A. A customized fixation plate with novel structure designed by topological optimization for mandibular angle fracture based on finite element analysis. *Biomed. Eng. Online* **2017**, *16*, 131. [CrossRef]

22. Planmeca ProMax 3D Max—Dental Imaging to the Max. Available online: https://www.planmeca.com/imaging/3d-imaging/planmeca-promax-3d-max/ (accessed on 09 April 2019).

23. Arango-Ospina, M.; Cortés-Rodriguez, C.J. Engineering design and manufacturing of custom craniofacial implants. In *The 15th International Conference on Biomedical Engineering*; Goh, J., Ed.; Springer International Publishing: Basel, Switzerland, 2014; pp. 908–911.

24. Zhou, L.; Shang, H.; He, L.; Bo, B.; Liu, G.; Liu, Y.; Zhao, J. Accurate Reconstruction of Discontinuous Mandible Using a Reverse Engineering/Computer-Aided Design/Rapid Prototyping Technique: A Preliminary Clinical Study. *J. Oral Maxillofac. Surg.* **2010**, *68*, 2115–2121. [CrossRef]

25. Van Bael, S.; Chai, Y.C.; Truscello, S.; Moesen, M.; Kerckhofs, G.; Van Oosterwyck, H.; Kruth, J.-P.; Schrooten, J. The effect of pore geometry on the in vitro biological behavior of human periosteum-derived cells seeded on selective laser-melted Ti6Al4V bone scaffolds. *Acta Biomater.* **2012**, *8*, 2824–2834. [CrossRef] [PubMed]

26. Ran, Q.; Yang, W.; Hu, Y.; Shen, X.; Yu, Y.; Xiang, Y.; Cai, K. Osteogenesis of 3D printed porous Ti6Al4V implants with different pore sizes. *J. Mech. Behav. Biomed. Mater.* **2018**, *84*, 1–11. [CrossRef] [PubMed]

27. Schaller, A.; Voigt, C.; Huempfner-Hierl, H.; Hemprich, A.; Hierl, T. Transient finite element analysis of a traumatic fracture of the zygomatic bone caused by a head collision. *Int. J. Oral Maxillofac. Surg.* **2012**, *41*, 66–73. [CrossRef]

28. El-Anwar, M.I.; Mohammed, M.S. Comparison between two low profile attachments for implant mandibular overdentures. *J. Genet. Eng. Biotechnol.* **2014**, *12*, 45–53. [CrossRef]

29. Ti6Al4V ELI Titanium Alloy. 2014. Available online: http://www.arcam.com/wp-content/uploads/Arcam-Ti6Al4V-ELI-Titanium-Alloy.pdf (accessed on 27 January 2019).

30. Szucs, A.; Bujtár, P.; Sándor, G.K.B.; Barabás, J. Finite element analysis of the human mandible to assess the effect of removing an impacted third molar. *J. Can. Dent. Assoc.* **2010**, *76*, a72.

31. Simonovics, J.; Bujtár, P.; Váradi, K. Effect of preloading on lower jaw implant. *Biomech. Hungarica* **2013**, *6*, 21–28. [CrossRef]

32. What is FDM?: Fused Deposition Modeling Technology for 3D Printing | Stratasys n.d. Available online: https://www.stratasys.com/fdm-technology (accessed on 6 January 2019).

33. High Resolution SLA and SLS 3D Printers for Professionals. Formlabs n.d. Available online: https://formlabs.com/ (accessed on 7 February 2019).

34. FDM vs SLA: How does 3D Printing Technology Work? |. Pinshape 3D Printing Blog | Tutorials, Contests & Downloads 2017. Available online: https://pinshape.com/blog/fdm-vs-sla-how-does-3d-printer-tech-work/ (accessed on 7 February 2019).

35. Kumar, A.; Nune, K.C.; Murr, L.E.; Misra, R.D.K. Biocompatibility and mechanical behaviour of three-dimensional scaffolds for biomedical devices: Process–structure–property paradigm. *Int. Mater. Rev.* **2016**, *61*, 20–45. [CrossRef]

36. Horn, T.J.; Harrysson, O.L.A.; Marcellin-Little, D.J.; West, H.A.; Lascelles, B.D.X.; Aman, R. Flexural properties of Ti6Al4V rhombic dodecahedron open cellular structures fabricated with electron beam melting. *Addit. Manuf.* **2014**, *1–4*, 2–11. [CrossRef]

37. Murr, L.E.; Gaytan, S.M.; Medina, F.; Lopez, H.; Martinez, E.; Machado, B.I.; Hernandez, D.H.; Martinez, L.; Lopez, M.I.; Wicker, R.B.; Bracke, J. Next-generation biomedical implants using additive manufacturing of complex cellular and functional mesh arrays. *Philos. Trans. R. Soc. A: Math. Phys. Eng. Sci.* **2010**, *368*, 1999–2032. [CrossRef]

38. Tang, H.P.; Wang, J.; Song, C.N.; Liu, N.; Jia, L.; Elambasseril, J.; Qian, M. Microstructure, mechanical properties, and flatness of sebm Ti-6Al-4V sheet in as-built and hot isostatically pressed conditions. *JOM* **2017**, *69*, 466–471. [CrossRef]

39. Wang, M.; Li, H.Q.; Lou, D.J.; Qin, C.X.; Jiang, J.; Fang, X.Y.; Guo, Y.B. Microstructure anisotropy and its implication in mechanical properties of biomedical titanium alloy processed by electron beam melting. *Mater. Sci. Eng. A* **2019**, *743*, 123–137. [CrossRef]

40. Balza, J.C.; Zujur, D.; Gil, L.; Subero, R.; Dominguez, E.; Delvasto, P.; Alvarez, J. Sandblasting as a surface modification technique on titanium alloys for biomedical applications: abrasive particle behavior. *IOP Conf. Ser. Mater. Sci. Eng.* **2013**, *45*, 012004. [CrossRef]

41. Geomagic Control X. 3D Systems n.d. Available online: https://www.3dsystems.com/software/geomagic-control-x (accessed on 10 February 2019).

42. Hammad Mian, S.; Abdul Mannan, M.; M. Al-Ahmari, A. The influence of surface topology on the quality of the point cloud data acquired with laser line scanning probe. *Sens. Rev.* **2014**, *34*, 255–265. [CrossRef]

43. Ridtzwan, M.I.; Solehuddin, S.; Hassan, A.Y.; Shokri, A.A.; Mohamad Ibrahim, M.N. Problem of Stress Shielding and Improvement to the Hip Implant Designs: A Review. *J. Med. Sci.* **2007**, *7*, 460–467.

44. Soh, C.-K.; Yang, Y.; Bhalla, S. (Eds.) *Smart Materials in Structural Health Monitoring, Control and Biomechanics*; Springer-Verlag: Berlin, Germany, 2012.

45. Basciftci, F.A.; Korkmaz, H.H.; Üşümez, S.; Eraslan, O. Biomechanical evaluation of chincup treatment with various force vectors. *Am. J. Orthod. Dentofac. Orthop.* **2008**, *134*, 773–781. [CrossRef] [PubMed]

Article

Diffraction Line Profile Analysis of 3D Wedge Samples of Ti-6Al-4V Fabricated Using Four Different Additive Manufacturing Processes

Ryan Cottam [†], Suresh Palanisamy [1,2,*], Maxim Avdeev [3], Tom Jarvis [4], Chad Henry [5], Dominic Cuiuri [6], Levente Balogh [7] and Rizwan Abdul Rahman Rashid [1,2]

[1] School of Engineering, Faculty of Science, Engineering and Technology, Swinburne University of Technology, Hawthorn, VIC 3122, Australia; rrahmanrashid@swin.edu.au
[2] Defence Materials Technology Centre, Hawthorn, VIC 3122, Australia
[3] The Bragg Institute, Australian Nuclear Science and Technology Organisation (ANSTO), Lucas Heights, NSW 2234, Australia; maxim.avdeev@ansto.gov.au
[4] Monash Centre for Additive Manufacturing, Monash University, Notting Hill, VIC 3168, Australia; tom.jarvis@monash.edu
[5] Commonwealth Scientific and Industrial Research Organization (CSIRO), Clayton, VIC 3168, Australia; wchadry@yahoo.com
[6] School of Mechanical, Materials, and Mechatronic Engineering, Faculty of Engineering and Information Sciences, University of Wollongong, Wollongong, NSW 2522, Australia; dominic@uow.edu.au
[7] Department of Mechanical and Materials Engineering, Queen's University, Kingston, ON K7L 3N6, Canada; levente.balogh@queensu.ca
*   Correspondence: spalanisamy@swin.edu.au; Tel.: +61-3-9214-5037
†   Deceased—12th January 2017.

Received: 27 November 2018; Accepted: 7 January 2019; Published: 9 January 2019

Abstract: Wedge-shaped samples were manufactured by four different Additive Manufacturing (AM) processes, namely selective laser melting (SLM), electron beam melting (EBM), direct metal deposition (DMD), and wire and arc additive manufacturing (WAAM), using Ti-6Al-4V as the feed material. A high-resolution powder diffractometer was used to measure the diffraction patterns of the samples whilst rotated about two axes to collect detected neutrons from all possible lattice planes. The diffraction pattern of a LaB_6 standard powder sample was also measured to characterize the instrumental broadening and peak shapes necessary for the Diffraction Line Profile Analysis. The line profile analysis was conducted using the extended Convolution Multiple Whole Profile (eCMWP) procedure. Once analyzed, it was found that there was significant variation in the dislocation densities between the SLMed and the EBMed samples, although having a similar manufacturing technique. While the samples fabricated via WAAM and the DMD processes showed almost similar dislocation densities, they were, however, different in comparison to the other two AM processes, as expected. The hexagonal (HCP) crystal structure of the predominant α-Ti phase allowed a breakdown of the percentage of the Burgers' vectors possible for this crystal structure. All four techniques exhibited different combinations of the three possible Burgers' vectors, and these differences were attributed to the variation in the cooling rates experienced by the parts fabricated using these AM processes.

Keywords: Ti-6Al-4V; additive manufacturing; selective laser melting (SLM); electron beam melting (EBM); direct metal deposition (DMD); wire and arc additive manufacturing (WAAM); diffraction line profile analysis; extended convolution multiple whole profile (eCMWP)

1. Introduction

Additive Manufacturing (AM) of metallic materials is receiving increasing attention worldwide [1–3]. There are two main AM approaches, and they are the powder bed approach, and the direct deposition approach. In the powder bed approach, a layer of powder is swept over a platform and the powder is melted together using either a laser or an electron beam, known as selective laser melting (SLM) and electron beam melting (EBM), respectively. The platform is then lowered and a new layer of powder is swept over. The melting process is then performed again according to the G-code given by the pre-processing software. This is repeated until the desired part is formed [4,5]. The direct deposition approach is used to melt either powder blown onto the substrate or wire fed into the melt pool of the substrate, using a heat source that is usually either a high powered laser or an electric arc, known as direct metal deposition (DMD) and wire arc additive manufacturing (WAAM), respectively. Tracks of material are placed side-by-side, layer-upon-layer, until the desired shape is formed. The powder bed approach easily produces complex shapes, with the aid of support structures, while the direct deposition approach produces more basic shapes but can produce much larger sized components because it is not limited by the size of the powder bed chamber [6].

In this study, the titanium alloy Ti-6Al-4V has been investigated, which has been widely used in aerospace and medical applications. Four leading metallic AM processes, namely SLM, EBM, DMD, and WAAM, were employed to manufacture wedge-shaped Ti-6Al-4V samples. While all four of these techniques are relatively mature technologies and have proven capable to build 3D shapes, the metallurgical character of the deposits, in particular differences between the different technologies, is not very well known and may play a role in identifying which of the technologies should be employed for the manufacturing of a particular component. One aspect of the metallurgical character of the deposits is the dislocation content, which has a significant influence on the mechanical properties of the deposits, in particular the strength and ductility. This knowledge aids in understanding why the mechanical properties of these AM processes varies, as that reported by Sames et al. [7] and Frazier [8]. Therefore, diffraction line profile analysis was employed to measure the dislocation contents produced by these AM technologies.

Diffraction line profile analysis (DLPA) is a diffraction analysis technique where the number and type of dislocations present in a structure can be determined quantitatively, together with other microstructural features, such as average sub-grain size, planar fault frequency, and the breakdown of the Burgers' vectors of the different dislocations and their relative percentages [9,10]. Moreover, DLPA is an indirect method used to derive average microstructural characteristics from neutron or X-ray diffraction patterns. In the case of neutron diffraction, the volume of the diffracting material is in the range of cubic centimetres, which allows a non-destructive and bulk characterization of the microstructure of the investigated material. It is most commonly used during plastic deformation, in parallel with plasticity models to understand the various slip system activities [11,12]. Using neutron DLPA, it has been shown that the initial dislocation density of as-built stainless steel samples depends on the type and the parameters of the applied AM process and can be altered by subsequent heat treatments or plastic deformation [13,14]. Results strongly suggest that the flow strength of the as-built AM stainless steel is primarily controlled by the dislocation density present in the material, making DLPA a useful characterization tool for such materials [13]. This type of analysis, which provides quantitative characteristics on the dislocation structure, deepens the understanding of the deformation mechanisms operating in metals and their correlation with the mechanical properties of the bulk polycrystal. This capability can be particularly useful for materials having hexagonal close-packed (HCP) crystal structures, as that of titanium, which possess anisotropic properties, and their dislocation structures vary with deformation temperature and grain orientation. The main source of dislocations in the metal AM structures form during the martensitic phase transformation occurring due to rapid solidification of the molten metal. As the crystal lattice changes from one crystal structure to another—in the case of Ti-6Al-4V, from β-Ti (HCP crystal structure) to α'-Ti (BCC crystal structure) due to displacive transformation—dislocations are formed at the transformation interface to allow for the misfit between the two crystal lattices, HCP and BCC [15]. Ahmed and Rack [16] have shown that the crystallography of the two types of martensite that forms for Ti-6Al-4V is different and is dependent on the cooling rate. Therefore, the dislocation content is an indication as to the nature and extent of the martensitic phase transformation, which is why the diffraction line profile analysis is vital in understanding the effect of the various AM processes on the microstructure and mechanical properties of Ti-6Al-4V. Moreover, this type of analysis may elucidate this change in the martensite formed by the change in the percentage of dislocations formed, as different orientation relationships at the transforming interface will change the type and number of dislocations formed, as that reported by Carroll et al. [17].

The primary objective of this study is to measure the different dislocation densities that form during four prominent metal AM processes, namely SLM, EBM, DMD, and WAAM, when fabricating Ti-6Al-4V wedge-shaped samples. This titanium alloy has been widely used for both aerospace and medical applications. Hence, in this study, the samples were irradiated with neutrons and the diffraction data was collected and analyzed using DLPA technique.

2. Materials and Methods

2.1. Sample Preparation

A wedge-shaped sample, with dimensions shown in Figure 1, was fabricated using each of the four AM processes, namely SLM, EBM, DMD, and WAAM. The wedge geometry was chosen to manifest changes in the character of the builds as a function of section size (e.g., thinner sections may cool faster than the thicker sections), which may influence microstructure formation and hence the mechanical properties. The processing parameters for the four different techniques were different and were optimized in a separate study by the various research providers.

An EOSINT M280 machine (Electro Optical Systems EOS GmbH, Krailling, Germany) was used to fabricate the titanium wedge-shape sample in the horizontal orientation using the SLM process. Gas atomized Ti-6Al-4V powder with particle size up to 63 μm was used. The process parameters used were: laser power 280 W, scan speed 1200 mm/s, layer thickness 30 μm, and hatch spacing 140 μm.

An Arcam A1 machine (Arcam AB, Mölndal, Sweden) was used for the EBM process. The titanium sample was fabricated using Ti-6Al-4V ASTM Grade 23 powder (average particle size of 73.52 μm). The standard Arcam theme 3.2.121 (Arcam AB, Mölndal, Sweden) was employed, which had an acceleration voltage of 60 kV, beam current of 1–10 mA, beam spot size of 200 μm, speed factor of 98, scanning line offset of 0.1 mm, layer thickness of 50 μm, and preheating temperature of 730 °C.

A 5 kW Trumpf-POM machine was used for the DMD process. A 210 mm × 115 mm × 6 mm titanium base plate in the annealed condition and Ti-6Al-4V powder (average particle size of 60 μm) supplied by TLS Technik were used for deposition. The powder was delivered to the deposition area. The fabrication of the wedge-shaped sample was conducted in an argon and helium gas atmosphere to minimize oxygen contamination. A laser power of 1600 W, laser spot size of about 2.2 mm, laser head traverse speed of 60 mm/min, and powder feed rate of 4.3 g/min were employed for processing.

A gas tungsten arc welding (WAAM) process was used for fabricating the Ti-6Al-4V sample. A 250 mm × 100 mm × 12 mm titanium base plate and a 1 mm diameter hard-drawn Ti-6Al-4V wire were used for deposition. However, the thick track dimensions of the WAAM process made it impossible to produce the wedge sample to the required dimensional tolerances by using only the deposition process. Instead, a rectangular block of material of dimensions 25 mm × 55 mm × 12 mm was deposited and the wedge shape was subsequently produced by wire cutting and machining. A current of 140 Amps, travel speed of 152 mm/min, arc length of 3.5 mm, and wire feed speed of 1.34 m/min (0.28 kg/h) were used as processing parameters. Welding grade argon (99.995% purity) was used as shielding gas at a flow rate of 25 L/min.

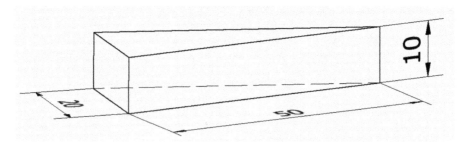

Figure 1. Dimensions of the wedge-shaped sample (in mm) used for the diffraction line profile analysis.

2.2. Neutron Diffraction Set Parameters and Extended Convolution Multiple Whole Profile (eCMWP) Analysis

Neutron diffraction data for the Ti-6Al-4V samples was collected on the high-resolution diffractometer Echidna at the OPAL facility (ANSTO, Lucas Heights, Australia), using neutrons of wavelength 1.6215 Å. During data collection, each sample was rotated around its two-fold axis to reduce the effect of preferred orientation, if any. To determine instrumental contribution to the peak width, calibration data was collected for a powder LaB_6 sample (NIST SRM 660c). The entire sample was simultaneously exposed to the neutron beam during the rotations in both orientations.

The analysis in this study was performed using the peak line broadening analysis software package eCMWP ("extended Convolution Multiple Whole Profile" procedure, 2017, G. Ribárik, et al., Budapest, Germany) [9,18]. The eCMWP software constructs a theoretical diffraction pattern based on well-established physical models of the microstructure, e.g., sub-grain size distributions, dislocation structures, and planar faults. The final shape of the various peaks of a diffraction pattern, $I^{PM}(2\theta)$, is a convolution of the contribution of various lattice defects and the contribution of the diffraction instrument itself, which is calculated using Equation (1) [9,18].

$$I^{PM}(2\theta) = \sum_{hkl} I^S_{hkl} * I^D_{hkl} * I^{PD}_{hkl} * I^{INST}_{hkl} + I_{BG} \tag{1}$$

where the defect related profile functions are the size, I^S_{hkl}, represents the dislocation cell or sub-grain size distribution [19], I^D_{hkl}, represents the contribution of the dislocations [20,21], I^{PD}_{hkl}, represents the contribution of the planar defects such as twin boundaries or stacking faults [22,23], I^{INST}_{hkl}, is the instrumental peak broadening and shape [24], and I_{BG} is the background of the diffraction pattern usually represented by a cubic spline. The instrumental peak shapes, I^{INST}_{hkl}, were determined by measuring a LaB_6 standard powder sample, which has no detectable microstructure (i.e., it is coarse grained, strain and dislocation free), thereby generating a result which is indicative of the peak broadening caused by the diffraction instrument itself. The resulting theoretical diffraction pattern presented in Equation (1) is fitted to the experimental data using a least-squares algorithm [18]. The fitting variables of the theoretical diffraction pattern are the quantitative features of the microstructure, such as the median and width of the sub-grain size distribution, density, type, and arrangement of the dislocations, and frequency of planar faults. The eCMWP software determines these quantitative characteristics by forward modelling the microstructure until a match is found between the theoretical and measured whole diffraction patterns [9,18].

Figure 2 shows the eCMWP refinement for the SLM sample. The open circles represent the measured data, the continuous line represents the modeled pattern, and the difference between the two is also shown. The 004, 202, and 104 reflections were not included in the analysis due to their low signal-to-noise ratio. Even though 200 and 210 also have low signal-to-noise ratios they were included, because ignoring them would have decreased the quality of the modeling for the overlapping high-intensity reflections. The eCMWP software matches the measured and the modeled full pattern by refining the microstructural parameters. In the present case, the refined microstructural parameters were: area weighted average sub-grain size $<X>_A$, total dislocation density ρ, and ratio of sub-densities having $<a>$, $<c+a>$, and $<c>$ Burgers' vector type, as described by Máthis et al. [10] and Ungár et al. [25]. It is important to note that the area weighted average sub-grain size $<X>_A$ represents a domain size which is defined by low-angle grain boundaries or dislocation walls, thus it will be referred to as sub-grain size or dislocation cell size; as reported by Ungár et al. [19]; hence, it is not the grain size defined by high angle grain boundaries visible in an optical microscope or in a low resolution TEM.

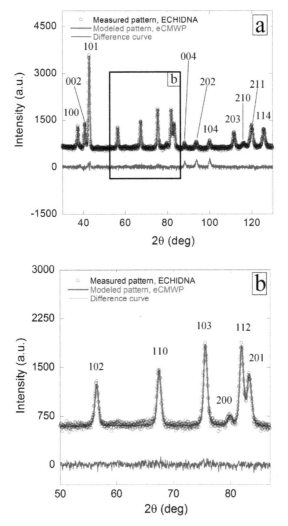

Figure 2. The eCMWP analysis, showing: (**a**) comparison of the measured and modeled pattern for the SLMed Ti-6Al-4V sample; and (**b**) diffraction plots between 50–85° (2θ) consisting of 102, 110, 103, 200, 112, and 201 phase reflections.

In order to qualitatively assess the neutron diffraction measurements, a Williamson-Hall (WH) plot was constructed. The WH plot presents the full width at half maximum (FWHM) of the peaks as a function of peak position. The FWHM values have been corrected for the instrumental broadening, thus, they represent only the broadening induced by the microstructural features found in the samples.

2.3. Residual Stress Analysis

Residual stress measurements were carried out using the contour method. The samples were wire-cut and surface profiles were measured on a Brown and Sharpe coordinate measuring machine (TESA USA, North Kingstown, RI, USA) equipped with a low force touch probe and 1 mm diameter ruby-tipped stylus. Each cut surface was measured with a 0.1 mm × 0.1 mm grid spacing, producing approximately 20,000 data points. The residual stresses were calculated from the raw contour data using MATLAB (Version 8.4, The Mathworks Inc., Natick, MA, USA) scripts and ABAQUS (Version 6.13, Dassault Systèmes Simulia Corp., Johnston, RI, USA) Finite Element code.

2.4. Microstructural Analysis

For microstructural characterization, the samples were cut from the mid-section of the wedge and prepared for metallographic examination. The samples were polished and etched with Kroll's reagent. The microstructures were examined under the Olympus BX-61 optical microscope (Olympus Corporation, Shinjuku, Japan).

3. Results and Discussion

The representative microstructures of the 3D printed titanium samples, fabricated using four different AM processes, are shown in Figure 3. It was noticed that all the samples consisted of α'-Ti martensitic phase along with $\alpha+\beta$-Ti matrix and prior β-Ti grain boundaries. However, the morphology of the α'-Ti martensite phase was slightly different, depending upon the AM process. The SLMed sample consisted of fine martensitic laths resembling a needle-like shape. The EBMed sample consisted of a similar type of fine martensitic laths, as that of the SLMed sample, but these laths were short and less in quantity. On the other hand, the DMDed sample consisted of thick and short martensitic laths with distinct prior β-Ti grain boundaries. The WAAMed sample consisted of a longer prior β-Ti grain boundary.

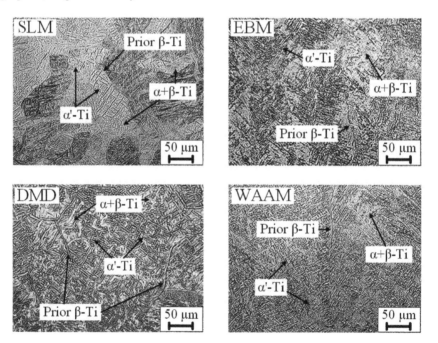

Figure 3. Microstructures of the 3D printed Ti-6Al-4V samples using four different AM processes.

The Williamson-Hall (WH) plot is shown in Figure 4, and the quantitative results obtained using eCMWP are presented in Table 1. From both the WH plot and the eCMWP data, it can be observed that the SLMed sample consisted of the highest dislocation density, whereas the EBMed sample had the lowest dislocation density in comparison to the DMDed and WAAMed samples, which exhibited similar dislocation densities. In contrast, the FWHM of the different diffraction peaks does not increase monotonously with increasing 'K', the reciprocal of the lattice spacing. This is due to the well-known effect of strain anisotropy [26], which is an indication of a significant dislocation density present in the material. The mathematical description of strain anisotropy is provided by the dislocation contrast factors, which describe the broadening of a given diffraction peak as a function of the hkl Miller indices and the different dislocation types. The evaluation of the contrast factors is handled internally by the eCMWP line profile analysis software, and the results can be used to determine the ratio of the dislocations densities having <*a*>, <*c+a*>, and <*c*> type Burgers' vectors [10].

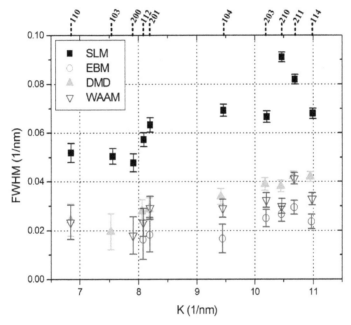

Figure 4. WH Plot representing the physical peak broadening as a function of peak position 'K'.

Table 1. Dislocation sub-cell sizes <X>_A, total dislocation densities ρ_{TOTAL}, and the ratios of sub-densities having <*a*>, <*c+a*>, and <*c*> Burgers' vector types for the Ti-6Al-4V specimen fabricated using four metal AM processes.

AM Process	$<X>_A$ (nm)	ρ_{TOTAL} (m^{-2})	<*a*> %	<*c+a*> %	<*c*> %
SLM	100 ± 15	$(24 \pm 3) \times 10^{14}$	85 ± 10	10 ± 10	5 ± 10
EBM	>500	$(1.4 \pm 0.5) \times 10^{14}$	70 ± 15	30 ± 15	0 ± 15
DMD	120 ± 20	$(4.1 \pm 0.5) \times 10^{14}$	60 ± 15	30 ± 15	10 ± 15
WAAM	>500	$(4.1 \pm 0.5) \times 10^{14}$	80 ± 10	20 ± 10	0 ± 10

From the eCMWP results presented in Table 1, it is quite evident that there is significant variation in the dislocation sizes and densities, which is process-dependent. It should be noted here that the entire sample was irradiated with neutrons simultaneously, and the values presented in Table 1 are an average across the sample. The WAAMed sample exhibited a large sub-grain size that can be attributed to the high heat input of the gas tungsten arc-welding (GTAW) torch used for its fabrication.

While not measured in this study, if the temperature is below the martensite finish temperature but still elevated during processing, it is possible that recovery can occur, thus increasing the sub-grain size [27]. This is further supported by the evidence that the WAAMed sample consisted of long prior β-Ti grain boundaries, which is possible when the cooling rates are low [28]. Likewise, the EBMed sample also has a high sub-grain size, which can be attributed to the high bed temperature, and subsequent slow cooling after deposition, allowing recovery processes to occur (i.e., the temperature allows the dislocations to annihilate each other). Therefore, the EBMed sample consisted of fewer α'-Ti martensitic laths (shown in Figure 3) compared to other samples [29]. On the other hand, the DMDed and SLMed samples exhibit smaller sub-cell sizes that indicates that the cooling rate during the process was higher than WAAM process, as well as the recovery rate being significantly reduced. However, the heat input in the DMD process was considerably higher than the SLM process, which was retained in the sample longer, resulting in thicker martensitic laths, as observed in the microstructures of these samples [15].

The EBMed sample has very low dislocation density, which can be the result of the high powder bed temperature (730 °C) employed during the EBM process. This provides sufficiently large amount of energy to drive recovery of any dislocations that form due the martensitic transformation [27]. However, the dislocation density in the SLMed sample is ~6× larger than in the WAAMed and DMDed samples and ~20× larger than the EBMed sample. This large increase in the dislocation content for the SLMed sample is not evident, given that these samples experience martensitic phase transformation during rapid cooling irrespective of the AM process used. There are three potential sources of the formation of dislocations during fabrication of these four samples, as follows; plastic deformation due to residual stress formation as the sample cools after deposition [30]; the dislocation formation of the displacive martensitic phase transformation [31]; and the plastic strain that the existing martensite laths undergo as new martensite laths form during the phase transformation [32]. As reported by Vasinonta et al. [30], the formation of residual stress in the 3D printing of Ti-6Al-4V is dependent on the process parameters. Therefore, this can be the first potential source of an increase in the dislocation content in the SLMed sample.

The residual stresses present in the Ti-6Al-4V samples along the central cross-section are shown in Figure 5. The SLMed sample contained a significant amount of compressive residual stress in the lower central portion of the wedge sample and tensile residual stress in the top and bottom edges. Such a steep residual stress gradient can result in the increased formation of dislocations in the sample [33,34]. In contrast, the EBMed sample had a more uniform residual stress state across the cross-section, ranging between −200 MPa and 200 MPa. The EBMed sample showed a much lower level of residual stress state compared to the SLMed sample, primarily due to the high powder bed temperature and the vacuum atmosphere maintained during the fabrication process [29]. The DMDed sample also consisted of uniform residual stresses, apart from a couple of pockets of compressive residual stresses and tensile residual stresses closer to the edges of the wedge sample, similar to that reported by Cottam et al. [35]. The WAAMed sample consisted of patches of mild tensile residual stress in the range of 100 to 300 MPa, dispersed throughout the cross-section of the wedge-shaped specimen owing to the much higher heat input than the laser-based DMD process [29,36].

The SLMed and WAAMed samples have approximately the same breakdown of dislocation types, from Table 1. The DMDed sample has a Burgers' vectors breakdown with more pyramidal, <c+a>, and prismatic (<c>) dislocations, whereas the EBMed sample consists of higher proportion of pyramidal, <c+a> type dislocations. Perhaps, this can be due to easier recovery of <a> type dislocations than the other two types. This is reasonable as the <a> type dislocations make up most of the dislocation content for Ti-6Al-4V samples fabricated using all four processes. Therefore, the probability that two <a> dislocations will meet and annihilate is higher than two <c+a> dislocations, resulting in an increase in the proportion of <c+a> type dislocations with increasing levels of recovery [27].

The crystallography of the strain associated with the martensitic phase transformation is invariant and as such the level of dislocations it introduces will be relatively consistent for the four 3D printing processes. The plastic strain of the existing martensitic laths as new martensitic laths form is dependent on the stress state of the material, as the martensitic transformation proceeds to a level which is a combination of the residual stress state and the local stress in the grain. Therefore, since the residual stress state of the SLMed sample is higher than for the WAAMed and DMDed samples, the plastic strain during the transformation will increase, and as a result, the amount of dislocations that will form in the SLMed sample will increase. Ali et al. [37] reported a decomposition of the α′-Ti martensite structure into a more homogeneous α+β phase upon preheating the SLM powder bed to about 570 °C. Furthermore, each of these AM processes experience different cooling rates, as graphically illustrated in Figure 6. Although not experimentally analyzed, it is expected that the cooling rate of the powder-bed metal AM processes is quite high, and for Ti-6Al-4V could be more than 525 °C/s, whereas for the other two AM processes, DMD and WAAM, the cooling rates are likely to be lower than SLM and EBM. Moreover, due to the high heat input in the WAAM process, the cooling rate can even be below 410 °C/s. Therefore, this explains the resultant microstructures, dislocation densities, and residual stresses of the titanium samples fabricated using these four different AM processes. It should be noted here that there will be process variations associated with each of the above AM processes which might yield a different result to that reported in this work. Therefore, further in-depth investigation is required to comprehend the dependency of the properties of the printed parts in terms of microstructure, phases, and dislocation densities on the process variables.

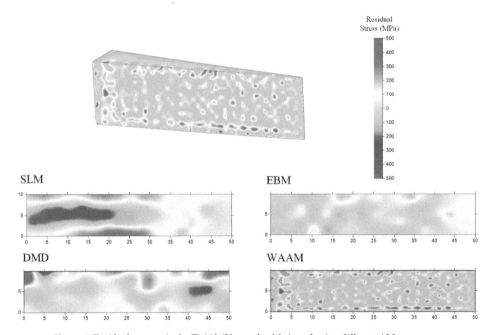

Figure 5. Residual stresses in the Ti-6Al-4V samples fabricated using different AM processes.

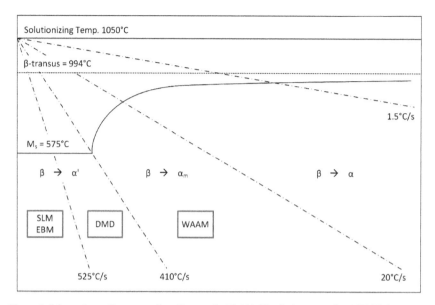

Figure 6. Schematic continuous cooling diagram for Ti-6Al-4V solution treated at 1050 °C for 30 min and quenched using the Jominey end quench test, showing the fit in terms of the cooling rates experienced during the four AM processes. (Reproduced with permission from Ahmed and Rack, Phase transformations during cooling in α+β titanium alloys, published by Elsevier, 1998 [16]).

4. Conclusions

Titanium alloys, such as Ti-6Al-4V, are widely used in aerospace and medical applications. In this work, the variation in the dislocation content of Ti-6Al-4V samples produced by the four metal AM processes, SLM, EBM, DMD, and WAAM, was investigated, and it was attributed to the different process characteristics. The SLMed sample contained a high dislocation content, the source of which was attributed to the volume of the martensitic phase transformation as well as the residual stresses in the sample. The DMDed sample had intermediate dislocation content and a significant amount of <c> and <c+a> dislocations due to the intermediate cooling rate experienced by the sample during this process. The EBMed sample exhibited a low dislocation content and a large dislocation cell size, which was attributed to the high temperature of the powder bed of 730 °C during the printing process, which facilitated recovery of the <a> type dislocations, thereby resulting in a decrease in the α'-Ti martensite phase. The WAAMed sample produced a dislocation content similar to that of the DMDed sample, but the breakdown of the dislocation of Burgers' vectors was significantly different, which was attributed to the low cooling rate during the process.

Author Contributions: Conceptualization, R.C. and S.P.; data curation, R.C. and M.A.; formal analysis, M.A.; funding acquisition, S.P.; investigation, R.C. and M.A.; methodology, R.C.; project administration, S.P.; resources, S.P., M.A., T.J., C.H., and D.C.; supervision, S.P.; validation, M.A. and L.B.; writing—review and editing, R.A.R.R.

Funding: This research was funded by Defense Materials Technology Centre, Project 1.11.

Acknowledgments: This paper includes research that was supported by DMTC Limited (Australia). The authors have prepared this paper in accordance with the intellectual property rights granted to partners from the original DMTC project. Swinburne University of Technology would also like to thank Girish Thipperudrappa for operating the DMD during the manufacturing of the wedge-shaped sample.

Conflicts of Interest: The authors declare no conflict of interest.

Metals **2019**, *9*, 60

Notations

3D	Three Dimensional
AM	Additive Manufacturing
BCC	Body Centered-Cubic
EBM	Electron Beam Melting
eCMWP	extended Convolution Multiple Whole Profile
DLPA	Diffraction Line Profile Analysis
DMD	Direct Metal Deposition
FWHM	Full Width at Half Maximum
HCP	Hexagonal Close-Packed
$I^{PM}(2\theta)$	Convolution of diffraction peaks at 2θ diffraction angle
I^{S}_{hkl}	Dislocation cell/sub-grain size distribution at [hkl] crystal plane
I^{D}_{hkl}	Contribution of the dislocations at [hkl] crystal plane
I^{PD}_{hkl}	Contribution of the planar defects at [hkl] crystal plane
I^{INST}_{hkl}	Instrumental peak broadening and shape at [hkl] crystal plane
I_{BG}	Background of the diffraction pattern
K	Reciprocal of the lattice spacing
SLM	Selective Laser Melting
TEM	Transmission Electron Microscope
WAAM	Wire Arc Additive Manufacturing
WH	Williamson-Hall
$<X>_{A}$	Area weighted average sub-grain size
ρ_{Total}	Total dislocation density

References

1. Kruth, J.P.; Leu, M.C.; Nakagawa, T. Progress in additive manufacturing and rapid prototyping. *CIRP Ann. Manuf. Technol.* **1998**, *47*, 525–540. [CrossRef]
2. Attar, H.; Ehtemam-Haghighi, S.; Kent, D.; Dargusch, M.S. Recent developments and opportunities in additive manufacturing of titanium-based matrix composites: A review. *Int. J. Mach. Tools Manuf.* **2018**, *133*, 85–102. [CrossRef]
3. Rashid, R.; Masood, S.H.; Ruan, D.; Palanisamy, S.; Rahman Rashid, R.A.; Elambasseril, J.; Brandt, M. Effect of energy per layer on the anisotropy of selective laser melted AlSi12 aluminium alloy. *Addit. Manuf.* **2018**, *22*, 426–439. [CrossRef]
4. Ponnusamy, P.; Masood, S.H.; Palanisamy, S.; Rahman Rashid, R.A.; Ruan, D. Characterization of 17-4PH alloy processed by selective laser melting. *Mater. Today* **2017**, *4*, 8498–8506. [CrossRef]
5. Agius, D.; Kourousis, K.; Wallbrink, C. A Review of the As-Built SLM Ti-6Al-4V Mechanical Properties towards Achieving Fatigue Resistant Designs. *Metals* **2018**, *8*, 75. [CrossRef]
6. Hoye, N.; Cuiuri, D.; Rahman Rashid, R.A.; Palanisamy, S. Machining of GTAW additively manufactured Ti-6Al-4V structures. *Int. J. Adv. Manuf. Technol.* **2018**, *99*, 313–326. [CrossRef]
7. Sames, W.J.; List, F.A.; Pannala, S.; Dehoff, R.R.; Babu, S.S. The metallurgy and processing science of metal additive manufacturing. *Int. Mater. Rev.* **2016**, *61*, 315–360. [CrossRef]
8. Frazier, W.E. Metal additive manufacturing: A review. *J. Mater. Eng. Perform.* **2014**, *23*, 1917–1928. [CrossRef]
9. Ribárik, G.; Ungár, T. Characterization of the microstructure in random and textured polycrystals and single crystals by diffraction line profile analysis. *Mater. Sci. Eng. A* **2010**, *528*, 112–121. [CrossRef]
10. Máthis, K.; Nyilas, K.; Axt, A.; Dragomir-Cernatescu, I.; Ungár, T.; Lukáč, P. The evolution of non-basal dislocations as a function of deformation temperature in pure magnesium determined by X-ray diffraction. *Acta Mater.* **2004**, *52*, 2889–2894. [CrossRef]
11. Glavicic, M.G.; Salem, A.A.; Semiatin, S.L. X-ray line-broadening analysis of deformation mechanisms during rolling of commercial-purity titanium. *Acta Mater.* **2004**, *52*, 647–655. [CrossRef]
12. Glavicic, M.G.; Semiatin, S.L. X-ray line-broadening investigation of deformation during hot rolling of Ti–6Al–4V with a colony-alpha microstructure. *Acta Mater.* **2006**, *54*, 5337–5347. [CrossRef]

13. Brown, D.W.; Adams, D.P.; Balogh, L.; Carpenter, J.S.; Clausen, B.; King, G.; Reedlunn, B.; Palmer, T.A.; Maguire, M.C.; Vogel, S.C. In Situ Neutron Diffraction Study of the Influence of Microstructure on the Mechanical Response of Additively Manufactured 304L Stainless Steel. *Metall. Mater. Trans. A Phys. Metall. Mater. Sci.* **2017**, *48*, 6055–6069. [CrossRef]

14. Pokharel, R.; Balogh, L.; Brown, D.W.; Clausen, B.; Gray, G.T., III; Livescu, V.; Vogel, S.C.; Takajo, S. Signatures of the unique microstructure of additively manufactured steel observed via diffraction. *Scr. Mater.* **2018**, *155*, 16–20. [CrossRef]

15. Rahman Rashid, R.A.; Palanisamy, S.; Attar, H.; Bermingham, M.; Dargusch, M.S. Metallurgical features of direct laser-deposited Ti6Al4V with trace boron. *J. Manuf. Process.* **2018**, *35*, 651–656. [CrossRef]

16. Ahmed, T.; Rack, H.J. Phase transformations during cooling in α+β titanium alloys. *Mater. Sci. Eng. A* **1998**, *243*, 206–211. [CrossRef]

17. Carroll, B.E.; Palmer, T.A.; Beese, A.M. Anisotropic tensile behavior of Ti–6Al–4V components fabricated with directed energy deposition additive manufacturing. *Acta Mater.* **2015**, *87*, 309–320. [CrossRef]

18. Ribarik, G.; Ungar, T.; Gubicza, J. MWP-fit: a program for multiple whole-profile fitting of diffraction peak profiles by ab initio theoretical functions. *J. Appl. Crystallogr.* **2001**, *34*, 669–676. [CrossRef]

19. Ungár, T.; Tichy, G.; Gubicza, J.; Hellmig, R. Correlation between subgrains and coherently scattering domains. *Powder Diffr.* **2005**, *20*, 366–375. [CrossRef]

20. Ungár, T.; Tichy, G. The Effect of Dislocation Contrast on X-Ray Line Profiles in Untextured Polycrystals. *Phys. Status Solidi* **1999**, *171*, 425–434. [CrossRef]

21. Borbély, A.; Ungár, T. X-ray line profiles analysis of plastically deformed metals. *C. R. Phys.* **2012**, *13*, 293–306. [CrossRef]

22. Balogh, L.; Ribárik, G.; Ungár, T. Stacking faults and twin boundaries in fcc crystals determined by x-ray diffraction profile analysis. *J. Appl. Phys.* **2006**, *100*, 023512. [CrossRef]

23. Balogh, L.; Tichy, G.; Ungár, T. Twinning on pyramidal planes in hexagonal close packed crystals determined along with other defects by X-ray line profile analysis. *J. Appl. Crystallogr.* **2009**, *42*, 580–591. [CrossRef]

24. Stokes, A.R. A numerical Fourier-analysis method for the correction of widths and shapes of lines on X-ray powder photographs. *Proc. Phys. Soc.* **1948**, *61*, 382. [CrossRef]

25. Ungár, T.; Balogh, L.; Ribárik, G. Defect-Related Physical-Profile-Based X-Ray and Neutron Line Profile Analysis. *Metall. Mater. Trans. A* **2010**, *41*, 1202–1209. [CrossRef]

26. Ungár, T. Dislocation model of strain anisotropy. *Powder Diffr.* **2008**, *23*, 125–132. [CrossRef]

27. Humphreys, F.J.; Hatherly, M. Chapter 6 - Recovery After Deformation. In *Recrystallization and Related Annealing Phenomena*, 2nd ed.; Humphreys, F.J., Hatherly, M., Eds.; Elsevier: Oxford, UK, 2004; pp. 169–213.

28. Dąbrowski, R. The Kinetics of Phase Transformations During Continuous Cooling of the Ti6Al4V Alloy from the Single-Phase β Range. *Arch. Metall. Mater.* **2011**, *56*, 703. [CrossRef]

29. Li, C.; Liu, Z.Y.; Fang, X.Y.; Guo, Y.B. Residual Stress in Metal Additive Manufacturing. *Proc. CIRP* **2018**, *71*, 348–353. [CrossRef]

30. Vasinonta, A.; Beuth, J.L.; Griffith, M. Process Maps for Predicting Residual Stress and Melt Pool Size in the Laser-Based Fabrication of Thin-Walled Structures. *J. Manuf. Sci. Eng.* **2006**, *129*, 101–109. [CrossRef]

31. Bhadeshia, H.K.D.H. Developments in martensitic and bainitic steels: Role of the shape deformation. *Mater. Sci. Eng. A* **2004**, *378*, 34–39. [CrossRef]

32. Christien, F.; Telling, M.T.F.; Knight, K.S. Neutron diffraction in situ monitoring of the dislocation density during martensitic transformation in a stainless steel. *Scr. Mater.* **2013**, *68*, 506–509. [CrossRef]

33. Mishurova, T.; Cabeza, S.; Artzt, K.; Haubrich, J.; Klaus, M.; Genzel, C.; Requena, G.; Bruno, G. An Assessment of Subsurface Residual Stress Analysis in SLM Ti-6Al-4V. *Materials* **2017**, *10*, 348. [CrossRef] [PubMed]

34. Yadroitsev, I.; Yadroitsava, I. Evaluation of residual stress in stainless steel 316L and Ti6Al4V samples produced by selective laser melting. *Virtual Phys. Prototyp.* **2015**, *10*, 67–76. [CrossRef]

35. Cottam, R.; Thorogood, K.; Lui, Q.; Wong, Y.C.; Brandt, M. The Effect of Laser Cladding Deposition Rate on Residual Stress Formation in Ti-6Al-4V Clad Layers. *Key Eng. Mater.* **2012**, *520*, 309–313. [CrossRef]

36. Szost, B.A.; Terzi, S.; Martina, F.; Boisselier, D.; Prytuliak, A.; Pirling, T.; Hofmann, M.; Jarvis, D.J. A comparative study of additive manufacturing techniques: Residual stress and microstructural analysis of CLAD and WAAM printed Ti–6Al–4V components. *Mater. Des.* **2016**, *89*, 559–567. [CrossRef]

37. Ali, H.; Ma, L.; Ghadbeigi, H.; Mumtaz, K. In-situ residual stress reduction, martensitic decomposition and mechanical properties enhancement through high temperature powder bed pre-heating of Selective Laser Melted Ti6Al4V. *Mater. Sci. Eng. A* **2017**, *695*, 211–220. [CrossRef]

Article

Fatigue Behavior of Non-Optimized Laser-Cut Medical Grade Ti-6Al-4V-ELI Sheets and the Effects of Mechanical Post-Processing

André Reck [1,*], André Till Zeuner [1] and Martina Zimmermann [1,2]

1   Institute of Materials Science, Technical University of Dresden, 01062 Dresden, Germany
2   Department of Materials Characterization and Testing, Fraunhofer-Institute for Material and Beam Technology IWS, 01277 Dresden, Germany
*   Correspondence: andre.reck@tu-dresden.de

Received: 18 June 2019; Accepted: 26 July 2019; Published: 30 July 2019

Abstract: The study presented investigates the fatigue strength of the ($\alpha+\beta$) Ti-6Al-4V-ELI titanium alloy processed by laser cutting with and without mechanical post-processing. The surface quality and possible notch effects as a consequence of non-optimized intermediate cutting parameters are characterized and evaluated. The microstructural changes in the heat-affected zone (HAZ) are documented in detail and compared to samples with a mechanically post-processed (barrel grinding, mechanical polishing) surface condition. The obtained results show a significant increase ($\approx$50%) in fatigue strength due to mechanical post-processing correlating with decreased surface roughness and minimized notch effects when compared to the surface quality of the non-optimized laser cutting. The martensitic α'-phase is detected in the HAZ with the formation of distinctive zones compared to the initial equiaxial $\alpha+\beta$ microstructure. The HAZ could be removed up to 50% by means of barrel grinding and up to 100% through mechanical polishing. A fracture analysis revealed that the fatigue cracks always initiate on the laser-cut edges in the as-cut surface condition, which could be assigned to an irregular macro and micro-notch relief. However, the typical characteristics of the non-optimized laser cutting process (melting drops and significant higher surface roughness) lead to early fatigue failure. The fatigue cracks solely started from the micro-notches of the surface relief and not from the dross. As a consequence, the fatigue properties are dominated by these notches, which lead to significant scatter, as well as decreased fatigue strength compared to the surface conditions with mechanical finishing and better surface quality. With optimized laser-cutting conditions, HAZ will be minimized, and surface roughness strongly decreased, which will lead to significantly improved fatigue strength.

Keywords: Titanium alloys; Ti-6Al-4V-ELI; fatigue; laser cutting; post-processing; α'-martensite; HAZ; barrel grinding; notch; fracture

---

1. Introduction

Titanium alloys are a frequently used material in industrial applications with ongoing market growth in recent years [1]. The industrial applications involve a wide spectrum from the aerospace and automobile sector, the chemical industry, to the field of medical engineering, such as applications in osteosynthesis. The reasons for this broad range of applications are high specific strength even at higher temperatures, excellent corrosion resistance, and the ability to adjust the material properties to a great extent with the optimization of the microstructure, as well as surface properties [1–3].

Titanium alloys possess high sensitivity to processing conditions and the subsequent fatigue loading during the application, strongly depending on the chemical composition, initial microstructure, and thermomechanical history [1,4–6]. Insufficient understanding or negligence of these factors may, therefore, result in an under- or overestimation of the fatigue strength and, as a consequence, either the failure to explore the full potential of titanium alloys or the leading to fatal failure cases.

The processing method of laser-cutting is a frequently used technique to transfer the sheet pre-product to the end geometry for the application. The high cost-effectiveness, economic efficiency, variability, as well as the possibility to produce complex geometries in a short time, are only a few of the many advantages [7–9]. However, the thermal input due to the laser must be adjusted and optimized regarding the surface roughness and resulting heat-affected zone (HAZ). Otherwise, micro- and macro-notch effects can negatively affect the fatigue behavior leading to catastrophic pre-mature failure in their application [4,10–12]. The local changes of the microstructure in the HAZ may also have a significant influence on the mechanical properties and on the fatigue strength, in particular [4,13–15]. However, the mechanical post-processing or finishing methods can improve the surface quality after laser-cutting or related processing methods in a significant way, which was proven in several studies [9,10,16–18]. The effects of surface roughness and the influence of the HAZ of a laser-cut component are, nevertheless, the subject of current research. Depending on the extent of strength reduction, the laser-cutting process has to be adjusted so as to form a favorable surface quality at the expense of the processing speed, while costly post-processing of the surface to remove the HAZ also has to be discussed with regard to its effectiveness on fatigue strength improvement.

The ($\alpha+\beta$) alloy Ti-6Al-4V-ELI, which is analyzed in the present study, is one of the standard alloys in the medical sector due to its exceptional combination of high specific strength, good ductility, and remarkable corrosion resistance. As sheet metal, this alloy is mostly processed by laser-cutting and mechanically post-processed with barrel or vibratory grinding methods. While detailed investigations on the surface quality and the HAZ due to laser-cutting and their consequence with regard to the fatigue properties are limited, it is of a common consensus that from a general point of view, a significant negative influence of increasing surface roughness on fatigue behavior is expected [10,14–16,19]. Studies on the general interaction of laser and material surface concentrate on the temperature field, kerf development, and parameter studies (type of laser, power, speed, etc.) [13,20–24]. Fatigue properties of Ti-6Al-4V-ELI have been investigated intensively in the past [10,14–17,25–29]. The sensitivity against surface roughness and underlying surface near microstructure is assessed and confirmed. Da Silva et al. [10] demonstrated a theoretical and experimental decrease in fatigue strength with increasing surface roughness. Morita et al. [27] focused on the influence of short term aging to improve the fatigue performance of notched Ti-6Al-4V-ELI. The development of α'-martensite phase during quenching led to a retardation of crack propagation and, therefore, better fatigue properties [27]. The mechanical post-processing methods to improve the fatigue properties of Ti-6Al-4V-ELI were the subject of investigations, as was the influence of the environment, which is especially important for application in the human body [10,16,17,30,31]. However, to the best of the author's knowledge, no direct studies are found focusing on the interaction between laser-cutting, the subsequent mechanical post-processing, and the endurable stress amplitudes. Furthermore, it is known from related studies concerning laser welding and comparable methods to which extent the laser can change the local microstructure, and therefore, the mechanical and fatigue properties [32–34].

Therefore, the study aims to identify the principle changes that are obtained regarding surface quality and near-surface microstructure due to laser-cutting with intermediate (non-optimized) cutting parameters and to evaluate these effects with respect to the fatigue behavior of Ti-6AL-4V-ELI. The possible local changes in the HAZ-microstructure shall be clarified, and the resulting surface quality assessed. Furthermore, the fatigue strength of the specimens after mechanical post-processing with the method of barrel grinding will be compared to samples with an as-cut surface and with a surface after manual mechanical polishing. Surface roughness is also investigated in conjunction with the crack initiation sites after fatigue loading.

2. Materials and Methods

The studied medical grade Ti-6Al-4V-ELI (ISO 5832-3) sheet pre-product was purchased from the supplier MetSuisse Distribution AG (Zug, Switzerland) and originally produced by RTI International Metals Inc. (Pittsburgh, PA, USA). The sheet thickness was 0.8 mm and the laser-cutting process was realized with a disk laser (TruDisk 5001-Fa. TRUMPF, Ditzingen, Germany) with 3 kW laser power at 25 m/min cutting speed. To minimize the possible chemical reactions due to oxygen or nitrogen, which can lead to hard and brittle TiO_2 or TiN surface layers, laser-cutting was carried out under argon atmosphere with an argon pressure of 6 bar. All laser cutting parameters (Table 1) were chosen to display an average parameter set for the Ti-6Al-4V-ELI alloy. Since the process optimization was not foreseen in the scope of the study presented, the parameters of the laser-cutting process were chosen on the basis of practical knowledge and represent an intermediate condition; however, allowing a general analysis of the possible changes in the microstructure and the extent of surface roughness effects on the fatigue behavior. The geometry used for the laser-cut fatigue samples was developed in previous studies on medical implant alloys [12] and is depicted in Figure 1.

Table 1. Laser cutting parameters applied for Ti-6l-4V-ELI sheets on the TruDisk 5001 laser (Fa. TRUMPF-Series Tru Laser 7025).

Cutting Parameters	Laser Power	Cutting Speed	Spot Size	Laser Beam Quality	Cutting Gas	Nozzle	Focal Distance
Used parameter set	3 kW ($\lambda = 1035$ nm)	25 m/min	150 µm (Focal spot on the surface)	$M^2 = 14$	Argon (6 bar)	Single head-Conical (2.0 mm)	6 inch

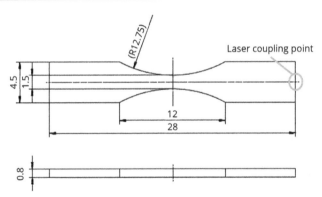

Figure 1. Fatigue sample geometry of Ti-6Al-4V-ELI with a marked laser coupling point. Dimensions in mm.

The samples originally processed by laser-cutting were divided into three series for fatigue testing. Laser-cut samples (Ti-6Al_LC), barrel-grinded samples (Ti-6Al_BG), and mechanically polished samples (Ti-6Al_MP) were investigated in detail before fatigue testing to document and assess the surface quality and near surface microstructure (heat-affected zone—HAZ) after the laser-cutting process, as well as after the mechanical post-processing. Therefore, metallographic cross-sections were prepared and microstructure analysis was executed by means of light microscopy, as well as scanning electron microscopy (SEM–JEOL JSM 7800-JEOL Ltd., Tokyo, Japan). The surface relief and roughness after laser-cutting was evaluated by confocal microscopy (Leica DCM 3D - Leica Microsystems, Wetzlar, Germany) using the software Leica Map Premium 6.2.7487. The element analysis for microstructural mapping in the HAZ was realized by energy dispersive spectroscopy (EDS–Oxford Instruments, Abingdon, Great Britain), which was attached on the SEM column. The AZtecHKL software (version 3.0) was used for evaluation.

All fatigue tests were carried out by a servo-electric load-frame (SEL 010 with software Loadframe SX 2.4–Fa. Thelkin, Winterthur, Switzerland) using an attached cooling device with pressurized air to hold the sample temperature during the fatigue tests at a constant room temperature level. The load-controlled tests were executed under tension-compression mode ($R = -1$) with a frequency of 20 Hz. The fatigue life limit N_G was set to 2×10^6 cycles. To ensure homogeneous sine oscillations, several pre-tests were carried out to find the best parameter sets of the test system for the specific vibration behavior of the Ti-6Al-4V-ELI alloy. Subsequently, load-increase tests were executed to find the suitable load horizons. Finally, series testing was realized by means of the staircase method (15–18 samples per series).

The Ti-6Al_LC samples were tested with the original surface condition after laser-cutting. The melting drops were removed in the clamping range to ensure safe testing. The Ti-6Al_BG samples were additionally barrel grinded with standard titanium alloy parameter sets (4 h (rough grit) and 4 h (fine grit) at 1000 RPM, pyramidal polishing stones) in a vibratory grinding machine for small samples (Fa. Rösler, Untermerzbach, Germany). The Ti-6Al_MP samples were manually mechanically polished along the cyclic load direction stepwise from P400, P800; P1200 to a final grid size of P2500, which corresponds to an average roughness value of Sa < 0.500 μm. The mechanical-polishing and barrel-grinding process were, thereby, carried out under constant fluid (water for mechanical polishing; grinding fluid for barrel (vibratory) grinding). Metallographic cross-sectioning documented the change in microstructure and HAZ as a consequence of the mechanical post-processing.

The fracture surface analysis was executed after the fatigue tests with SEM to locate crack initiation sites of all samples and to assess the possible connection between surface roughness-related notch effects due to laser-cutting and actual crack initiation after fatigue failure.

3. Results and Discussion

3.1. Surface Quality

The surface quality achieved due to the laser-cutting process with the particular cutting parameters applied (see Material and Methods section) is depicted in Figures 2 and 3. The Ti-6Al-4V-ELI alloy shows a distinctive surface relief with recognizable additional melting drops on the lower cutting edge. On the upper cutting edge, an irregular distributed terrace-like surface structure can also be identified (Figure 3a). Both observed features can be traced back to the influence of the laser-cutting parameters applied. The melting drops are caused by an insufficient pressure of the inert argon gas atmosphere. Whereas on the upper cutting edge, the occurring melting drops could be sufficiently removed, the lower cutting edge is less accessible, demanding a higher pressurized air stream to successfully remove all residual melting drops. As a consequence, the necessary argon pressure has to be well over 6 bar to homogeneously remove melting drops on the whole cutting edge.

The irregular terrace-like structure can be assigned to an influence throughout the interaction of the laser and Ti-6Al-4V-ELI but not directly explained till now. Further evaluation is, therefore, necessary. However, optimized laser-cutting parameters will improve the surface quality to a great extent, and therefore, also has a positive effect on the resulting fatigue strength of the Ti-6Al-4V-ELI alloy. The average height parameters of the laser-cut surface relief are measured according to ISO 25178 by means of confocal microscopy. Several μm in distance between the relief hills and valleys (Figure 2a) are identified, which concludes a significant surface roughness. Since fatigue failure in the LCF ($< 1 \times 10^3$–10^4 cycles) and HCF (up to 1×10^7 cycles) regime is mostly triggered by the stress concentration at distinct surface flaws leading to early crack initiation and propagation, pronounced surface roughness is detrimental for the expected fatigue strength and endurable cycles [10–12]. The Ti-6Al_LC samples with a surface quality as depicted in Figure 2 could, therefore, be expected to cause a dominant concentration of cyclic stress in the relief valleys during fatigue. In addition, Ti-6Al-4V is known for its high notch sensitivity, which was the subject of detailed studies in the past [3,10,14,19], demonstrating a strong decrease of fatigue strength correlating with higher surface roughness and resulting notch factors. Comparing the surface quality after laser-cutting (Ti-6Al_LC) with the sample series of Ti-6Al-BG and Ti-6Al_MP, the latter ones show a significantly improved surface quality, which correlates also with lower surface roughness. The mechanically polished surfaces of the Ti-6Al_MP samples are completely free of visible micro-notches which is a result of the homogeneous polishing to a grit size of P2500 and corresponds to a surface quality of $S_a < 0.500$ μm. In comparison to the as-cut surface, as well as the barrel grinded surface, this post-processing represents the best surface quality for the fatigue samples in the study presented.

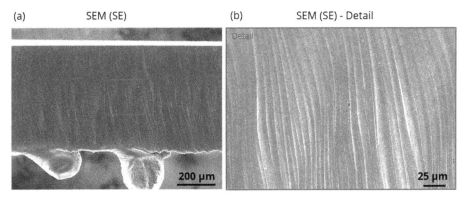

Figure 2. Obtained quality of the laser-cut surfaces of Ti-6Al-4V-ELI with non-optimized laser cutting conditions: (**a**) Overview SEM-image of the cutting edge with residual melting drops at the lower part; (**b**) Detailed SEM-image showing a wavy, irregular surface structure after the laser-cutting process.

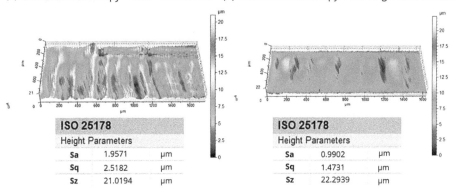

Figure 3. Typical representative surface reliefs by means of confocal microscopy (normalized) of Ti-6Al-4V-ELI with average height parameters: (**a**) Surface after laser-cutting; (**b**) Surface after barrel-grinding—Improvement of surface quality.

Barrel-grinded surfaces, on the other hand, show a strong improvement in surface quality and less roughness, but retained micro-notches are visible in the center area of the cutting edges (Figure 3b). These micro-notches could act as possible crack initiation sites in the application, which was already shown in detail in one of the authors previous studies of laser-cut β-titanium, as well as α-titanium for fatigue samples and osteosynthesis plates [12]. This leads to the conclusion that the barrel-grinding process is an effective tool to reduce detrimental laser-cutting effects, but only under the prerequisite of already optimized laser-cutting parameters for a specific material. Without the improvement of the original laser-cut surface quality, micro-notches seem to remain in the center area of the laser-cutting edges playing a decisive role for achievable fatigue strength in the application.

3.2. Microstructural Development of the Heat-Affected Zone (HAZ)

The microstructural response to the thermal input from processing in the heat-affected zone is alongside the surface quality, the most important factor to define the influence of the laser-cutting process regarding the fatigue behavior. Possible microstructural changes compared to the initial $\alpha+\beta$-microstructure could be expected for Ti-6Al-4V-ELI since the laser-cutting temperature lies well above the β-transus temperature of this alloy ($\approx$980 °C [35]). Hence, the laser-cutting process could change the surface-near microstructure caused by the local heating and subsequent self-quenching.

The HAZ in a cross-sectional view at the upper cutting edge depicted in Figure 4 clearly demonstrates the change in microstructure from equiaxial $\alpha+\beta$ into acicular martensitic α' with retained β, the latter being proven by the EDS-analysis showing a higher V-content as a strong β-phase stabilizer. The transformation into martensitic α' is, thereby, associated with the self-quenching process, which is fast enough to transform the initial equiaxial α-phase (5–10 μm in grain size) to fine acicular α' martensite. The retained β-phase can be explained by the very fast processing time and holding over β-transus not allowing the β-phase also to transform into α' martensite. The lower cutting edge with residual melting drops (Figure 2a) shows, by contrast, two distinctive zones (Figure 5—Cross-sectional view). A small surface layer (average thickness 10–15 μm) consists solely of acicular α' without β-phase. The second zone, which develops with growing distance to the free surface, consists of α' and β-phase, comparable to the upper cutting edge, followed by the initial ($\alpha+\beta$)-microstructure. The development of the different microstructural zones could be explained with the self-quenching gradient from the laser-cutting temperature, which causes different quenching speeds towards the sample interior and, therefore, possible zone formation. However, in order to explain the pronounced microstructural difference of upper and lower cutting edges, more detailed information of the temperature distribution related to the process parameters would be necessary.

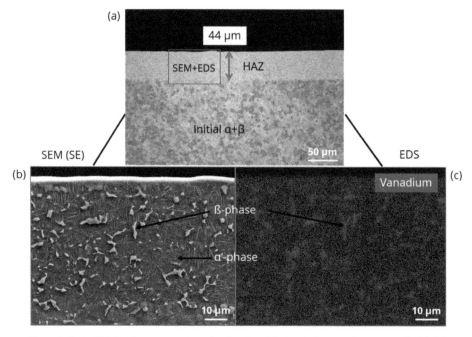

Figure 4. Typical HAZ at the upper cutting edge developed during the laser cutting process (Ti-6Al_LC sample): (**a**) Light microscopic image representing an overview; (**b**) SEM-image (SE cross-section) of the HAZ microstructure with α'-martensite; (**c**) Element analysis (EDS) showing the higher vanadium content (β-stabilizer) in the β-phase of the HAZ.

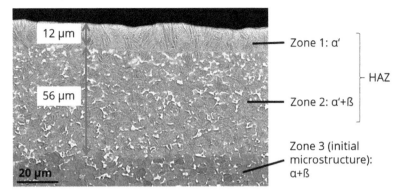

Figure 5. SEM image (BSE cross-section) of the HAZ at the lower cutting edge of a Ti-6Al_LC sample showing three distinctive microstructural zones after laser-cutting consisting of martensitic α'-phase and retained β-phase.

The measured thickness of the HAZ ($\approx$40–70 μm) depends on the occurrence of the different zones and is more pronounced at the lower cutting edge with the additional α'-zone. The optimized laser-cutting parameters, such as speed and power would contribute to a significant less pronounced HAZ. Furthermore, the application of a pulsed laser is expected to have a positive effect on the local microstructural changes due to the very short processing window and heating over the β-transus temperature [36].

With the mechanical post-processing of the laser-cut surface, not only the surface quality is improved, as was discussed in Section 3.1, but also the HAZ is significantly affected. The result is depicted in Figure 6 showing a significant reduction of the HAZ thickness. The barrel-grinding process creates, thereby, a decrease of about 50% HAZ compared to the initially observed HAZ without any mechanical post-processing. The reduction is independent of upper or lower cutting edge but shows irregularities at the crossover region of the radius and clamping range. This phenomenon can be explained by the mechanism of the barrel grinding process itself, which is based on a rotary or vibratory grinding process between the sample and a specific grinding stone, as well as an abrasive medium [37,38]. The careful selection and adaption of this process is a prerequisite for an optimal and homogeneous grinding result. The complex geometries and very poor initial surface quality are prerequisites that may have an adverse effect on the grinding quality. Insufficient grinding results negatively influence the fatigue properties with residual micro-notches on geometrically inaccessible locations on samples or components [12].

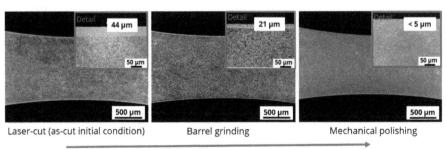

Figure 6. Thickness of the HAZ dependent on the mechanical post-processing showing a significant decrease in measured HAZ-thickness at the upper cutting edge.

The results of mechanical polishing are also depicted in Figure 6. An almost complete removal of the HAZ is observed. However, it has to be pointed out that in the study presented, the mechanical polishing was executed hand-made, leading to slight variations regarding the extent of removal of the HAZ for the overall number of samples.

3.3. Fatigue Results and Crack Initiation

The fatigue test results for all the three sample series are shown in Figure 7a. Remarkable differences are recognizable between the samples with the as-cut surface (Ti-6Al_LC) and the sample series with additional post-processing of the surface (Ti-6Al_BG and Ti-6Al_MP). The results for the as-cut samples show a significant scatter and the lowest fatigue strength ($\sigma_{aD,50\%}$ = 235 MPa) for a fatigue life limit of $N_G = 2 \times 10^6$. A superior fatigue strength was reached for the series with barrel-grinded surface (351 MPa) and mechanically polished surface (363 MPa). These values represent an increase of around 50% compared to the original laser-cut surface condition. However, while comparing the fatigue strengths obtained for the different sample conditions, it has to be considered that the as-cut condition represents a more or less "worst-case" condition since the process parameters were chosen on the basis of practical knowledge and were not yet optimized. The optimized laser-cutting parameters result in strongly decreased surface roughness and improved fatigue strength. The scattering of the results is less distinctive with mechanical post-processing of the surface and can be attributed to a higher surface quality with less surface roughness, and therefore, less possible notches acting as stress concentrators during fatigue loading. For the barrel grinded surface condition; however, micro-notches remain in the center area of the laser-cut edges, which was exemplarily shown in Figure 3. In comparison to the process of mechanical polishing, which displays the best overall fatigue behavior, barrel grinding of the Ti-6Al-4V-ELI alloy must be optimized, especially in case of homogeneity and removal of all surface notches.

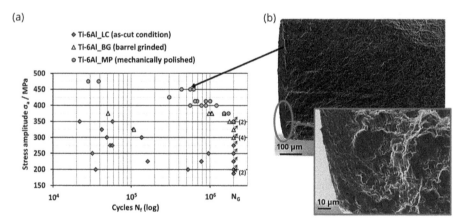

Figure 7. Fatigue results of the Ti-6Al-4V-ELI alloy: (**a**) S-N diagram for all three tested sample series; (**b**) Fracture surface analysis for a Ti-6Al_MP sample with mechanically polished surface conditions including the crack initiation site on the laser-cut edge directly near the sample corner after fatigue failure at 450 MPa and 6.2×10^5 cycles.

Fatigue crack initiation sites for all Ti-6Al-4V-ELI samples are located at the sample edges and corners. There are distinct differences between the three tested series, which are depicted in Figures 7b, 8 and 9. For the mechanically polished surface condition (Figure 7b), fatigue cracks initiate at the laser-cut edges or the flat sample edges, but primarily at or directly near the sample corners. The reason for this behavior is the overall good surface quality and comparable surface roughness of all sample edges with no distinctive micro-notches. In this case, the sample corners act normally as the highest stress concentrator under fatigue loading for flat sample geometries [39]. The consequence is a preferred fatigue crack initiation at these sites for the Ti-6Al_MP samples.

(a) Side view on laser-cut edge   (b) View on fractured surface

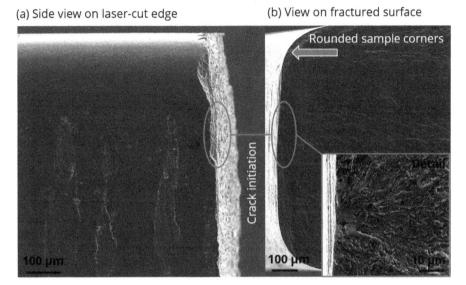

Figure 8. Fractography of a Ti-6Al_BG sample with barrel-grinded surface condition failed at 375 MPa and 5.1×10^4 cycles: (**a**) SEM (SE) analysis of the laser-cut edge with detectable notches; (**b**) The corresponding fracture surface with detailed view on fatigue crack initiation.

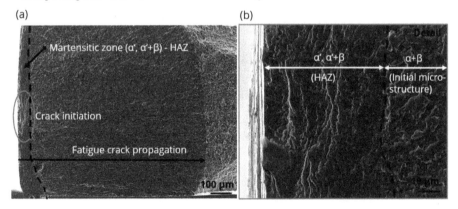

Figure 9. Fractography of a Ti-6Al_LC sample with as-cut surface condition failed at 200 MPa and 5.2×10^5 cycles: (**a**) SEM (SE) analysis of the fractured surface with identification of the martensitic zone; (**b**) The corresponding detailed view at the fatigue crack initiation point.

The sample series with the barrel-grinded surface condition (Ti-6Al_BG) showed, on the contrary, fatigue crack initiation always on the laser-cut edges, which is exemplarily depicted in Figure 8. Furthermore, the crack initiation sites are not always located at the smallest loaded cross-section but rather in the radius section. The cause for this behavior can be assigned to the remaining and detected surface notches after the barrel-grinding process in the middle of the laser-cut edges (Figures 3 and 8a). The irregular size and distribution of these remaining notches leads consequently to stress concentration, crack initiation, and failure at the biggest notches, and these can be located outside the smallest loaded cross-section. Associated with the effective lower overall surface roughness of the laser-cut edges and the observed, particularly rounded, sample corners (Figure 8b—lower stress concentration sites), all Ti-6Al_BG samples are expected to fail at one of the remaining notches.

The sample series with the initial as-cut surface condition (Ti-6Al_LC) reveal almost similar fatigue crack initiation points as the Ti-6Al_BG samples. Due to the significantly stronger local surface roughness (compared to the flat sample edges) caused by the laser-cutting for the process parameters applied, fatigue failure could always be directly assigned to the crack initiation at the laser-cut edges, which is depicted in Figure 9 in the fracture surface analysis. Pronounced irregularity of the surface profile leads subsequently to the high scattering of the fatigue results and the decreased fatigue strength for the Ti-6Al_LC sample series. At this point, it should be pointed out that, irrespective of the severe geometrical irregularity of the dross formation (see Figure 2a) due to a lack of process optimization, fatigue crack initiation always started from an extrusion/intrusion from the surface relief and not from a dross. The optimized cutting conditions improve this behavior and lead to significantly improved fatigue strength, although the formation of a slight surface relief cannot be completely avoided and may have an effect on the fatigue behavior. By comparing the fatigue behavior of the barrel-grinded and the as-cut condition, it becomes obvious that even marginal remains of the surface relief can result in early failure with values for the barrel grinded condition close to the upper scattering range of the as-cut results. The fatigue results obtained for the stress amplitudes between 300 MPa and 350 MPa demonstrate the particularly high sensitivity of the fatigue behavior of Ti-6Al-4V-ELI on surface flaws with early failures ($N < 2 \times 10^5$), as well as run-out samples ($N = 2 \times 10^6$) for the same stress level.

An aspect to be considered for all three sample series, but especially for the as-cut surface condition, is the presence of the HAZ caused by the laser-cutting. The martensitic zone (α' and $\alpha'+\beta$—Figures 4 and 5) is especially pronounced in the as-cut surface condition (Figure 6) and expected to influence the fatigue behavior, although an optimized laser-cutting process will significantly decrease HAZ development. Due to a higher strength and more interphase boundaries, this phase can have a positive effect on the fatigue strength or the fatigue crack growth behavior of Ti-6Al-4V [3,4,29]. However, in connection with higher or irregular surface roughness, α'-martensite can have an opposing effect and decrease the fatigue strength due to higher sensitivity for crack initiation and growth, which can be explained with higher hardness and lower ductility of α' compared with the initial $\alpha+\beta$ microstructure [16,19]. In the study presented, a pronounced HAZ can clearly be observed for all samples with an as-cut surface condition (Figure 9). The specific implication on the resulting fatigue strength seems; however, not significant due to the superior effect of the surface roughness, which is supported by the fact of highly scattered results for Ti-6Al_LC samples. In order to clearly identify an influence of the HAZ on the early stages of fatigue crack initiation and growth, further investigations with pre-notched samples would be necessary in order to quantitatively measure early crack growth rates in the HAZ. For the sample series with mechanical post-processing, HAZ was also observed, but in a less distinctive manner. The superior role of surface roughness is however confirmed with crack initiation on specific notches for the barrel grinded surface condition or near sample corners for the mechanically polished surface condition. It should be mentioned that possible residual stresses introduced in the near-surface area and their potential influence on the fatigue strength has not been part of the investigation of this study. However, this effect must be taken into consideration when further elucidating the superior fatigue behavior of the mechanically polished condition.

A comparison of the findings in the study presented with fatigue properties in the literature show a good correlation with the lower range of possible fatigue strengths for the ($\alpha+\beta$) Ti-6Al-4V alloy [14,15,25,26,29]. The overall fatigue properties of Ti-6Al-4V are, thereby, dependent on various influencial factors regarding the microstructure, the fatigue test conditions, as well as the fatigue geometry, composition and additional surface treatments. Furthermore, the role of thermal and residual stresses and their influence on the fatigue behavior must be taken into account [2–4,16,29]. Although the residual stress measurements were not executed in the study presented, it can be assumed that compressive residual stresses are introduced during the process of mechanical finishing. For the as-cut condition, the higher hardness of the α'-zone indicates an increase in mechanical strength; however, regarding the low surface quality of the as-cut condition, this general increase in strength can be assumed to be more susceptible for early fatigue failure and crack initiation due to an increased notch sensitivity. The compressive residual stresses introduced by means of mechanical polishing, on the other hand, contribute to an enhanced fatigue strength, and combined with an improved surface quality will result in an improved fatigue behavior [3,4,29]. The discrepancies in fatigue strength of Ti-6Al-4V_ELI in the study presented clearly correlates with the surface quality of each processing condition and, in particular, the resulting significant notch effects caused by the non-optimized laser-cutting process. In consequence, an optimization of the used laser-cutting parameters could be expected to significantly improve fatigue behavior.

The results of the as-cut surface condition impressively demonstrate the detrimental effect of the surface roughness with exceptional low fatigue strength ($\sigma_{aD,50\%}$ = 235 MPa) and pronounced scatter of the fatigue results. However, optimum laser-cutting parameters decrease this negative effect and strongly enhance fatigue behavior, because of the direct correlation of laser-cutting parameters and resulting surface quality. The barrel-grinding process (Ti-6Al_BG samples) results in a strong increase in fatigue strength, but is, nevertheless, determined by the dominating influence of the remaining notch profile in the center area of the laser-cut edges. The difference in fatigue strength compared to the mechanically polished surface condition (Ti-6Al_MP samples—Figure 7) can mainly be explained by the effect of the flat sample geometry and the consequential influence of the sample corners. These corners act as preferential, and possible early crack initiation sites in case of the Ti-6Al_MP samples with mechanically polished surface, whereas for Ti-6Al_BG samples with barrel grinded surface condition observed significant rounding of the corners (Figure 8) has the opposite effect. Hence, both sample series with mechanical post-processing of the surface exhibit almost similar fatigue behavior. The influence of the underlying microstructure and investigated HAZ seems to play only a circumstantial role compared to the dominant influence of the surface roughness. In consequence, the optimization of the surface roughness has to be the essential goal for laser cutting of Ti-6Al-4V-ELI in order to avoid early fatigue failure in application. Mechanical post-processing by means of barrel (vibratory) grinding or polishing seems to be an excellent choice for achieving this goal due to clear enhancement of fatigue strength caused by the minimization of arising surface roughness during laser cutting. Furthermore, residual compressive stresses could be introduced, positively influencing stress concentration and subsequent crack initiation during fatigue loading. All these factors are controlled by the grinding and polishing parameters, as well as the exact alloy composition, heat treatment condition, and original surface quality due to laser cutting, which, if mutually adjusted, hold the potential for significant fatigue life improvement. The mechanical post-processing, such as barrel grinding, can easily be implemented in industrial process chains and quality control. The downside is, on the other hand, the control of dimensions, especially in the case of maintaining the original edge proportion, which are prone to rounding (Figure 8b). Only recognition of all influencing effects can, therefore, lead to economic and positive post-processing with subsequent utilization of the full material potential under fatigue loading.

4. Conclusions

The study presented investigated the fatigue behavior of the medical implant alloy Ti-6Al-4V-ELI processed by laser-cutting and the consequences of additional surface post-processing on the fatigue properties with regard to surface roughness and HAZ. The fatigue behavior of the as-cut surface condition was compared with mechanically polished surfaces, as well as with barrel-grinded surfaces. The results can be summarized and concluded as follows:

- The surface relief introduced by the non-optimized laser-cutting influences the fatigue behavior of Ti-6Al-4V-ELI significantly. For the process parameters featured in this study, fatigue strength of the as-cut condition results in a drastic decrease of the fatigue strength compared to the additionally surface-treated condition. However, the difference in fatigue strength observed in this study will be controllable by optimizing the laser-cutting parameters.

- The main reason for the superior fatigue strength of the post-processed conditions is a minimized surface roughness, which in turn is responsible for higher resistance against fatigue crack initiation on macro- and micro-notches originally caused by the laser-cutting.

- The process of barrel-grinding after the laser-cutting was effective but revealed retained surface roughness in the center area of the cutting edges, which acts as preferred crack initiation sites compared to the particularly rounded sample corners.

- Mechanically polished samples always failed at or near the sample corners, which is caused by a stress concentration on these sites.

- The HAZ consisting of martensitic α' and β along distinctive surface and subsurface zones was analyzed and does not play a significant role in early fatigue failure, which instead was dominated by the surface roughness. Nevertheless, the applied mechanical post-processing led to an almost complete removal of the HAZ.

- To avoid early fatigue failure in the application, an optimization of the laser-cutting parameters is crucial in order to obtain better surface quality. This allows the required post-processing to improve the surface roughness further and, therefore, the fatigue strength. However, both processes, laser-cutting and mechanical post-processing, have to be optimized in the dependence of the specific alloy composition and fatigue behavior of Ti-6Al-4V-ELI.

Author Contributions: Conceptualization—A.R., M.Z.; Methodology—A.R.; Investigation—A.R., A.T.Z.; Data analysis/curation—A.R., A.T.Z.; Writing—original draft preparation—A.R.; Writing—review and editing—M.Z., A.R.; Project administration—A.R., M.Z.; Funding acquisition—M.Z.

Funding: This research was funded by Deutsche Forschungsgemeinschaft (DFG) under the grant number 402380578 (ZI 1006/14-1).

Acknowledgments: The authors thank Nikolai Schröder and the working group laser cutting at IWS for carrying out the laser cutting as well as Robert Kühne and Sebastian Schettler from the department Materials Characterization and Testing for helpful thematic discussions regarding the fatigue testing. Further thanks to Clemens Grahl for confocal microscopy as well as the metallographic team of the IWS, Lars Ewenz and Sebastian Schöne for sample preparation and assistance.

Conflicts of Interest: The authors declare no conflict of interest.

References

1. Banerjee, D.; Williams, J.C. Perspectives on titanium science and technology. *Acta Mater.* **2013**, *61*, 844–879. [CrossRef]
2. Leyens, C.; Peters, M. *Titanium and Titanium Alloys. Fundamentals and Applications*, 1st ed.; Wiley-VCH: Weinheim, Germany, 2003.
3. Lütjering, G.; Williams, J.C. *Titanium*, 2nd ed.; Springer: Berlin, Germany, 2007.
4. Bache, M. Processing titanium alloys for optimum fatigue performance. *Int. J. Fatigue* **1999**, *21*, 105–111. [CrossRef]
5. Chandravanshi, V.; Prasad, K.; Singh, V.; Bhattacharjee, A.; Kumar, V. Effects of $\alpha+\beta$ phase deformation on microstructure, fatigue and dwell fatigue behavior of a near alpha titanium alloy. *Int. J. Fatigue* **2016**, *91*, 100–109. [CrossRef]
6. Ezugwu, E.O.; Wang, Z.M. Titanium alloys and their machinability—A review. *J. Mater. Process. Technol.* **1997**, *68*, 262–274. [CrossRef]
7. Steen, W.M. *Laser Material Processing*, 3rd ed.; Springer: London, UK, 2003.
8. Davim, J.P. *Lasers in Manufacturing*; Wiley-VCH: Chichester, UK, 2013.
9. Yilbas, B.S. *The Laser Cutting Process. Analysis and Applications*; Elsevier Science: San Diego, CA, USA, 2017.
10. da Silva, P.S.C.P.; Campanelli, L.C.; Escobar Claros, C.A.; Ferreira, T.; Oliveira, D.P.; Bolfarini, C. Prediction of the surface finishing roughness effect on the fatigue resistance of Ti-6Al-4V ELI for implants applications. *Int. J. Fatigue* **2017**, *103*, 258–263. [CrossRef]
11. Pessoa, D.F.; Herwig, P.; Wetzig, A.; Zimmermann, M. Influence of surface condition due to laser beam cutting on the fatigue behavior of metastable austenitic stainless steel AISI 304. *Eng. Fract. Mech.* **2017**, *185*, 227–240. [CrossRef]

12. Reck, A.; Pilz, S.; Calin, M.; Gebert, A.; Zimmermann, M. Fatigue properties of a new generation ß-type Ti-Nb alloy for osteosynthesis with an industrial standard surface condition. *Int. J. Fatigue* **2017**, *103*, 147–156. [CrossRef]

13. Yang, J.; Sun, S.; Brandt, M.; Yan, W. Experimental investigation and 3D finite element prediction of the heat affected zone during laser assisted machining of Ti6Al4V alloy. *J. Mater. Process. Technol.* **2010**, *210*, 2215–2222. [CrossRef]

14. Eylon, D.; Pierce, C.M. Effect of microstructure on notch fatigue properties of Ti-6Al-4V. *Metall. Trans. A* **1976**, *7*, 111–121. [CrossRef]

15. Stráský, J.; Janeček, M.; Harcuba, P.; Bukovina, M.; Wagner, L. The effect of microstructure on fatigue performance of Ti-6Al-4V alloy after EDM surface treatment for application in orthopaedics. *J. Mech. Behav. Biomed. Mater.* **2011**, *4*, 1955–1962. [CrossRef]

16. Sonntag, R.; Reinders, J.; Gibmeier, J.; Kretzer, J.P. Fatigue performance of medical Ti6Al4V alloy after mechanical surface treatments. *PLoS ONE* **2015**, *10*, e0121963. [CrossRef] [PubMed]

17. Nalla, R.K.; Altenberger, I.; Noster, U.; Liu, G.Y.; Scholtes, B.; Ritchie, R.O. On the influence of mechanical surface treatments—deep rolling and laser shock peening—on the fatigue behavior of Ti–6Al–4V at ambient and elevated temperatures. *Mater. Sci. Eng. A* **2003**, *355*, 216–230. [CrossRef]

18. Nie, X.; He, W.; Zhou, L.; Li, Q.; Wang, X. Experiment investigation of laser shock peening on TC6 titanium alloy to improve high cycle fatigue performance. *Mater. Sci. Eng. A* **2014**, *594*, 161–167. [CrossRef]

19. Guilherme, A.S.; Henriques, G.E.P.; Zavanelli, R.A.; Mesquita, M.F. Surface roughness and fatigue performance of commercially pure titanium and Ti-6Al-4V alloy after different polishing protocols. *J. Prosthet. Dent.* **2005**, *93*, 378–385. [CrossRef]

20. Arif, A.F.M.; Yilbas, B.S. Thermal stress developed during the laser cutting process: Consideration of different materials. *Int. J. Adv. Manuf. Technol.* **2008**, *37*, 698–704. [CrossRef]

21. Sharma, A.; Yadava, V. Experimental analysis of Nd-YAG laser cutting of sheet materials—A review. *Opt. Laser Technol.* **2018**, *98*, 264–280. [CrossRef]

22. Sheng, P.S.; Joshi, V.S. Analysis of heat-affected zone formation for laser cutting of stainless steel. *J. Mater. Process. Technol.* **1995**, *53*, 879–892. [CrossRef]

23. Tamilarasan, A.; Rajamani, D. Multi-response optimization of Nd:YAG laser cutting parameters of Ti-6Al-4V superalloy sheet. *J. Mech. Sci. Technol.* **2017**, *31*, 813–821. [CrossRef]

24. Pandey, A.K.; Dubey, A.K. Modeling and optimization of kerf taper and surface roughness in laser cutting of titanium alloy sheet. *J. Mech. Sci. Technol.* **2013**, *27*, 2115–2124. [CrossRef]

25. Mower, T.M. Degradation of titanium 6Al–4V fatigue strength due to electrical discharge machining. *Int. J. Fatigue* **2014**, *64*, 84–96. [CrossRef]

26. Carrion, P.E.; Shamsaei, N.; Daniewicz, S.R.; Moser, R.D. Fatigue behavior of Ti-6Al-4V ELI including mean stress effects. *Int. J. Fatigue* **2017**, *99*, 87–100. [CrossRef]

27. Morita, T.; Tanaka, S.; Ninomiya, S. Improvement in fatigue strength of notched Ti-6Al-4V alloy by short-time heat treatment. *Mater. Sci. Eng. A* **2016**, *669*, 127–133. [CrossRef]

28. Akahori, T.; Niinomi, M. Fracture characteristics of fatigued Ti–6Al–4V ELI as an implant material. *Mater. Sci. Eng. A* **1998**, *243*, 237–243. [CrossRef]

29. Wu, G.Q.; Shi, C.L.; Sha, W.; Sha, A.X.; Jiang, H.R. Effect of microstructure on the fatigue properties of Ti–6Al–4V titanium alloys. *Mater. Des.* **2013**, *46*, 668–674. [CrossRef]

30. Papakyriacou, M. Effects of surface treatments on high cycle corrosion fatigue of metallic implant materials. *Inter. J. Fatigue* **2000**, *22*, 873–886. [CrossRef]

31. Roach, M.D.; Williamson, R.S.; Zardiackas, L.D. Comparison of the corrosion fatigue characteristics of CP Ti-Grade 4, Ti-6Al-4V ELI, Ti-6Al-7Nb, and Ti-15Mo. *J. ASTM Int.* **2005**, *2*, 12786. [CrossRef]

32. Gao, X.-L.; Zhang, L.-J.; Liu, J.; Zhang, J.-X. Porosity and microstructure in pulsed Nd:YAG laser welded Ti6Al4V sheet. *J. Mater. Process. Technol.* **2014**, *214*, 1316–1325. [CrossRef]

33. Hong, K.-M.; Shin, Y.C. Analysis of microstructure and mechanical properties change in laser welding of Ti6Al4V with a multiphysics prediction model. *J. Mater. Process. Technol.* **2016**, *237*, 420–429. [CrossRef]

34. Xu, P.-q.; Li, L.; Zhang, C. Microstructure characterization of laser welded Ti-6Al-4V fusion zones. *Mater. Charact.* **2014**, *87*, 179–185. [CrossRef]

35. Factory certification of RTI International Metals Inc., Ingot-No.: 9711830.

36. Shanjin, L.; Yang, W. An investigation of pulsed laser cutting of titanium alloy sheet. *Opt. Lasers Eng.* **2006**, *44*, 1067–1077. [CrossRef]
37. Rowe, W.B. *Principles of Modern Grinding Technology*; William Andrew: Oxford, UK, 2009.
38. Yang, S.; Li, W. *Surface Finishing Theory and New Technology*; Springer: Berlin, Germany, 2018.
39. Schijve, J. *Fatigue of Structures and Materials*; Springer: Dordrecht, The Netherlands, 2009.

Review

Evaluation of Fatigue Behavior in Dental Implants from In Vitro Clinical Tests: A Systematic Review

Rosa Rojo [1,*,†], María Prados-Privado [2,3,†], Antonio José Reinoso [4] and Juan Carlos Prados-Frutos [1]

1   Department of Medicine and Surgery, Faculty of Health Sciences, Rey Juan Carlos University, 28922 Alcorcon, Spain; juancarlos.prados@urjc.es
2   Department Continuum Mechanics and Structural Analysis Higher Polytechnic School, Carlos III University, 28911 Leganes, Spain; mprados@ing.uc3m.es
3   Asisa Dental (Engineering Researcher), José Abascal 32, 28003 Madrid, Spain
4   Department of ICT Engineering, Alfonso X El Sabio University, 28691 Madrid, Spain; areinpei@myuax.com
*   Correspondence: rosa.rojo@urjc.es; Tel.: +34-914-888-817
†   These authors contributed equally to this work.

Received: 18 March 2018; Accepted: 26 April 2018; Published: 3 May 2018

Abstract: In the area of dentistry, there is a wide variety of designs of dental implant and materials, especially titanium, which aims to avoid failures and increase their clinical durability. The purpose of this review was to evaluate fatigue behavior in different connections and implant materials, as well as their loading conditions and response to failure. In vitro tests under normal and dynamic loading conditions evaluating fatigue at implant and abutment connection were included. A search was conducted in PubMed, Scopus, and Science Direct. Data extraction was performed independently by two reviewers. The quality of selected studies was assessed using the Cochrane Handbook proposed by the tool for clinical trials. Nineteen studies were included. Fourteen studies had an unclear risk and five had high risk of bias. Due to the heterogeneity of the data and the evaluation of the quality of the studies, meta-analysis could not be performed. Evidence from this study suggests that both internal and morse taper connections presented a better behavior to failure. However, it is necessary to unify criteria in the methodological design of in vitro studies, following methodological guidelines and establishing conditions that allow the homogenization of designs in ISO (International Organization for Standardization) standards.

Keywords: biomechanics; dental implant(s); in vitro; systematic reviews; evidence-based medicine

1. Introduction

The use of dental implants has become a common practice for replacing missing teeth in different clinical situations [1,2]. The used materials are chosen according to both their mechanical and chemical properties, as well as to their biocompatibility [3]. Commercially, pure titanium and its alloys are widely used for manufacturing dental implants because of excellent mechanical and physical properties, and favorable rates for long-term clinical survival [4]. In addition, titanium-based implants have a good resistance to corrosion with an excellent biocompatibility and high modulus of elasticity [5,6]. The use of Ti–6Al–4V [7] alloy is employed for biomedical application and, also, in dental implants due to its high mechanical resistance, which ensures load transmission to bone tissues over a long time, which is necessary when damaged hard tissues are replaced by prostheses [3]. This alloy presents a drawback due to the use of vanadium and aluminum that can cause some toxic effects. Other titanium grades can be employed in dental implants but they also have disadvantages like the Young modulus, relatively low mechanical strength, poor wear resistance and difficulty to improve the mechanical properties without reducing biocompatibility. New β-type titanium alloys for dental implants have been developed. These have good properties and less toxicity, good ductility, high resistance. They also have elastic modules closer to those of human bone compared to other alloys [7].

Although the mechanical strength of dental implants is important, they must also present adequate stiffness to avoid shielding the bones from stress. This stress shielding induces loss of bone density, leading to bone atrophy. Moreover, the interaction between titanium and tissues is a key factor in the success of dental implants and, for this reason, surfaces of used alloys are conveniently treated. If the implant is manufactured with titanium grade 5, those implants must have a surface treatment to improve the corrosion [3]. This interaction between titanium and tissues is affected by the implant surface composition, as well as by its hydrophilicity, morphology, and roughness [8]. Different surface treatments have been tried and developed with the aim of obtaining titanium surfaces with better biological properties. This surface treatment yields a good osseointegration and obtain an improvement on the success of dental implants [9] with a change in the chemical composition. Nano roughness, texture, and porosity are some of the most important factors in the surface of an implant because they affect the ability of cells to adhere to a solid substrate [10].

The manufacturing process also influences the alloy's characteristics. The tensile strength of titanium alloy can range from 369 to 3267 MPa depending on the process employed. Fatigue behavior is also affected by the manufacturing process and it can be also improved by combining the material properties, surface properties, and design optimization of implants.

Despite all the advantages of titanium, considered as the "gold standard" material for the manufacture of dental implants, its biggest drawback are aesthetic considerations. Therefore, manufacturers began to use other types of materials such as ceramics [4,11] or polymers [4].

Osseointegrated dental implants are described in detail in a great number of studies [12], although implant-supported connection have been less disclosed in the literature [13].

Nowadays, a vast number of implant designs are available. The first dental implants had an external connection, where the hexagonal anti-rotational component is the most common design. Figure 1a shows an example of this connection. Due to the high rate of rejected implants, a new connection was designed. In this case, internal connection (Figure 1b) allowed a better union between the implant and the abutment. Finally, morse taper connection (Figure 1c), which is another option for an internal connection, was introduced because of its improvement on screw loosening [14–16].

These designs are different in terms of the connection of the implant-abutment (external or internal connections) [17,18], which from a mechanical point of view, is the weakest area of the implant system [19].

Applying static load tests to evaluate the strength of the implants and their components is a common practice. However, these tests do not simulate real situations for implants [2]. Considering that masticatory forces are cyclic, a fatigue testing should be carried out to predict how long an implant system is going to function properly [20]. In vitro tests should better simulate the clinical situations [21–23] and allow clinicians to understand the probability of survival of prosthetic components and implants [13]. Before implant components are launched to the market, they should satisfy the ISO 14801 specifications [24]. This ISO recommendation was planned for single, endosteal, transmucosal dental implants tested under worst case applications. Nevertheless, several testing protocols for evaluating the mechanical reliability of dental implants are available in the literature [25]. Different loading angles, frequency of loads, and application load levels have been employed in several published cyclic testing protocols [26].

The fracture of an implant or of any of its components is an important complication which limits the lifetime of the reconstruction. Although most of the studies available are limited to 5–7 years of follow-up [26], Snauwaert et al. found in a 15-year study, an early implant fracture (up to one year after abutment connection) in 3.4% and late implant failures of 7.4% [27]. Considering that implants should serve for decades, these type of studies are inadequate to analyze implant failure or fracture and make essential to examine dental implants under fatigue testing approaches. Once the number of cycles an implant can support until failure are known, its expected life can be predicted accurately [28].

The aim of this systematic review was to evaluate fatigue behavior in different connections and types of implants, their loading conditions, and their response to failure between implant and abutment.

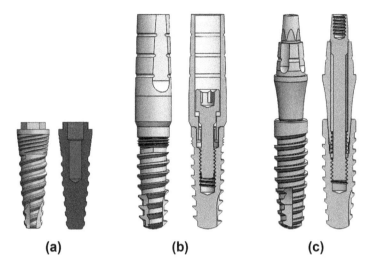

(a) **(b)** **(c)**

Figure 1. Schematic views of dental implant connections: (**a**) External connection (**b**) Internal connection (**c**) Morse taper connection.

2. Materials and Methods

This systematic review follows the Preferred Reporting Items for Systematic Reviews and Meta-Analyses (PRISMA) guidelines [29].

2.1. Focused Questions

First question (A): Does a certain number of cycles and a certain force (Newtons) exist between a defined implant connection and abutment failure?

Second question (B): How many cycles does a certain implant connection and abutment fail?

2.2. Inclusion and Exclusion Criteria

Inclusion criteria were in vitro clinical studies on dental implants in which fatigue at the implant and abutment connection were evaluated by subjecting them to dynamic cyclic loads. Studies were been carried out under normal environmental conditions. Dental implants are included disregarding the type of connection. There were no restrictions on the language or date of publication.

Exclusion criteria were all designs of observational studies, reviews, thermal fatigue assessment, static tests, or using incorrect units of measure.

2.3. Search Strategy

In this paper the research questions were elaborated considering each of the components of the PICO(S) [30] strategy research questions which is explained as follows: (P) dental implants and abutments; (I) cyclic loads; (C) studies with or without a comparison group where external, internal or morse taper connections were evaluated with implant materials and/or abutment of titanium, zirconium or others; (O) the evaluation of fatigue in terms of failure; (S) in vitro study.

An electronic search was performed in MEDLINE/PubMed, Scopus and Science Direct, database until the 15 March 2018. The search strategy used is detailed in Table 1.

Table 1. Search strategies carried out in databases.

Database	Search Strategy	Search Data
MEDLINE/PubMed	(dental AND (implant OR abutment) OR tooth implant) AND (cyclic loading) AND ((internal OR external) connection) AND (fatigue OR moment OR stress) AND ("in vitro" OR "experimental study") NOT (review)	15 March 2018
Scopus	(dental AND (implant OR abutment) OR tooth implant) AND (cyclic loading) AND ((internal OR external) connection) AND (fatigue OR moment OR stress) AND ("in vitro" OR "experimental study") AND NOT (review)	15 March 2018
Science Direct	(dental AND (implant OR abutment) OR tooth implant) AND (cyclic loading) AND ((internal OR external) connection) AND (fatigue OR moment OR stress) AND ("in vitro" OR "experimental study") AND NOT (review)	15 March 2018

2.4. Study Selection

Two authors (Rosa Rojo and María Prados-Privado) performed all the search operations and selected articles fulfilling the inclusion criteria independently and in duplicate. Additionally, the references of the articles included in this work were manually reviewed. Disagreements between the two authors were reviewed in a complete text by a third author (Juan Carlos Prados-Frutos) to make the final decision. The level of agreement between the reviewers regarding study inclusion was calculated using Cohen's kappa statistic.

2.5. Data Extraction

Two of the authors (María Prados-Privado and Rosa Rojo) collected all the data from the selected articles in duplicate and independently.

2.6. Study Quality Assessment

The assessment of risk of bias from clinical in vitro studies was evaluated by two of the authors (María Prados-Privado and Antonio José Reinoso), who were previously trained by an expert in evaluation of systematic reviews. For the assessment of risk of bias the Cochrane Handbook [31] was followed which incorporates seven domains: random sequence generation (selection bias); allocation concealment (selection bias); masking of participants and personnel (performance bias); masking of outcome assessment (detection bias); incomplete outcome data (attrition bias); selective reporting (reporting bias); and other bias.

The articles that did not achieve consensus between the two authors were reviewed by a third author (Rosa Rojo) to make the final decision.

The studies were classified into the following categories: low risk of bias—low risk of bias for all key domains; unclear risk of bias—unclear risk of bias for one or more key domains; high risk of bias—high risk of bias for one or more key domains.

2.7. Statistical Analysis

To evaluate the agreement between the inter-examiner, the statistic Cohen's kappa and the interpretation proposed by Landis & Koch [32] was used. Statistical calculations were performed with R software version 3.4.1 (R Core Development Team, R Foundation, Vienna, Austria) with the interrater reliability (irr) package.

3. Results

3.1. Study Selection

Figure 2 shows a flowchart of the study selection. All electronic search strategies provided 161 potential articles. Two of the authors (Rosa Rojo and María Prados-Privado) independently identified 48 eligible documents. The general agreement of eligibility of the studies between the authors was high ($k = 0.87$; $p = 0.049$.) Specifically, agreement between authors on the selection of articles in each considered database was high for all of them: Medline/PubMed ($k = 0.93$; $p = 0.046$), Scopus ($k = 0.80$; $p = 0.043$), and Science Direct ($k = 0.82$; $p = 0.05$). A total of 38 studies were excluded because they did not meet the defined inclusion criteria. Additionally, a manual search has been carried out to analyze the references cited in 10 of the articles that were included in this work. We reviewed 342 references. After removing duplicates, we analyzed the titles, abstracts and, when required, the full-text from 272 citations. A total of 263 studies were excluded as they did not match the inclusion criteria. As a result, nine additional articles were incorporated from the manual search. Finally, a total of nineteen in vitro studies were analyzed.

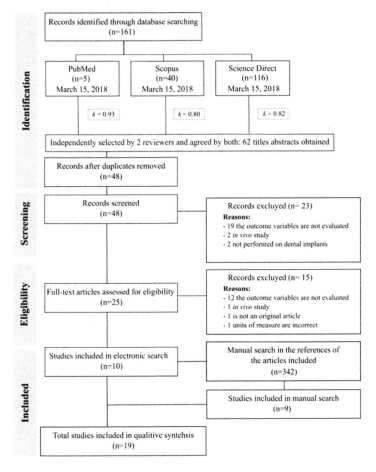

Figure 2. Flow chart.

3.2. Relevant Data from Studies

Two of the authors (María Prados-Privado and Rosa Rojo) extracted all the data from the selected works whose characteristics are shown in Table 2. The 19 in vitro studies analyzed [20,33–50] were carried out in several countries: Japan, Italy, Germany, Australia, Brazil, Switzerland, Turkey, Republic of Singapore, Greece, and the United States.

For all the studies, we collected the most important variables that could affect the results, such as sample size, the existence of funding, main characteristic of dental implants (connection, material, diameter and length), main properties of abutments (material and length), and applied load (magnitude, angulation, frequency and cycles).

Table 3 shows the 14 studies that answered the first question (A). The best behavior analyzed corresponded to a study where the zirconium-based implant hardened with alumina-doped yttrium-stabilized yttrium zirconium polyurethane and abutment of the same material was applied with a force of 98 N to 10,000,000 cycles [48]. No fault displayed.

Moreover, were are studies where the zirconium-based implant and abutment with an applied force of 50 N at 100 N, 45° of the axial axis of the dental implant and from 1,200,000 [46] to 3,600,000 cycles [50], respectively, showed failure under similar conditions.

It was observed that the titanium had a better behavior if the implant and the abutment were made of the same material [42,44,46,50]. The results also suggest that the behavior of the titanium worsens when the materials of the implant and the abutment are different [20,36,38].

Table 4 shows the six studies that answer the second question (B). The maximum number of reported cycles where the connection between the implant and the pillar fails is 5,000,000 cycles [33,41,45]. From the studies presenting better behavior fatigues, it has been found that the implant and the abutment are both made from titanium [33,45], or the abutment is combated with zirconium [41].

For both questions, Tables 3 and 4 show that the implant and the abutment behave better in the internal connections [33,41–46,48,50] and morse taper [41,47,49]. The study conducted by Mitsias et al. [47] answers the two research questions addressed in this work.

3.3. Study Quality Assessment

Evaluation of selection bias: They were only included in two of the analyzed studies of the method of randomization used [43,45]. However, it does not indicate whether there was concealment of this allocation.

Evaluation of performance bias: In all the studies analyzed there was no blinding of staff or assessors. Moreover, we found that in [34] the evaluator who prepared the specimens and who performed the tests were the same person. This fact may lead to a high risk of bias.

Assessment of detection bias: The results were not blinded in any of the studies.

Evaluation of attrition bias: All studies reported the complete results of the specimens defined in the clinical trial, although some reported inaccurately without indicating in the results the variable descriptions in their methodology [39,42].

Evaluation of notification bias: All studies provide detailed results with the exception of one, this study did not describe correctly whether the variables are quantitative or qualitative [39].

Evaluation of other bias: Funding was considered as another possible risk of bias in study designs. Eleven trials were funded by commercial firms [20,35,36,40–43,46,48–50], five were not reported [33,34,37,38,44], and three reported that no funding existed [39,45,47].

Using the evaluation of the seven domains for risk of bias it was determined that five had a high risk of bias [20,33–36], 14 an unclear risk [37–50], and none had a low risk of bias. Figure 3 shows a detailed description of the risk assessment of bias in the included studies.

Table 2. Main characteristics of the included studies.

Author/Year	Country	Journal	n	G	Financing
Balfour et al. [33] 1995	United States	Journal of Prosthetic Dentistry	21	3	U
Khraisat et al. [34] 2002	Japan	Journal of Prosthetic Dentistry	14	2	U
Çehreli et al. [35] 2004	Turkey	Clinical Oral Implants Research	8	1	Y
Butz et al. [36] 2005	United States	Journal of Oral Rehabilitation	48	3	Y
Gehrke et al. [37] 2006	Germany	Quintessence International	7	1	U
Kohal et al. [38] 2009	Germany	Clinical Implant Dentistry and Related Research,	48	3	U
Scarano et al. [39] 2010	Italy	Italian Oral Surgery	20	1	N
Magne et al. [40] 2011	Switzerland	Clinical Oral Implants Research	28	2	Y
Seetoh et al. [41] 2011	Republic of Singapore	The International Journal of Oral & Maxillofacial Implants	30	6	Y
Dittmer et al. [20] 2012	Germany	Journal of Prosthodontic Research	60	2	Y
Stimmelmayr et al. [42] 2012	Germany	Dental Materials	6	2	Y
Foong et al. [43] 2013	Australia	Journal of Prosthetic Dentistry	22	2	U
Pintinha et al. [44] 2013	Brazil	Journal of Prosthetic Dentistry	48	2	N
Marchetti et al. [45] 2014	Italy	Implant Dentistry	15	2	Y
Rosentritt et al. [46] 2014	Germany	Journal of Dentistry	64	8	N
Mitsias et al. [47] 2015	Greece	The International Journal of Prosthodontics	36	2	N
Spies et al. [48] 2016	Germany	Journal of the Mechanical Behavior of Biomedical Materials	48	3	Y
Guilherme et al. [49] 2016	United States	Journal of Prosthetic Dentistry	57	3	Y
Preis et al. [50] 2016	Germany	Dental Materials	60	6	Y

n: sample size; G: Number of groups; Y: Yes; N: No; U: Unclear.

Table 3. Variables analyzed from the answer to question A.

Author/Year	Implant				Abutment		Applied Load			Cycles	Failure
	Connection	Material	Diameter	Length	Material	Length	Magnitude (N)	Angulation (°)	Frequency (Hz)		
Çehreli et al. [35] 2004	-	-	10	-	-	-	75 ± 5	20	0.5	500,000	N
Butz et al. [36] 2005	E	-	4	13	Ti	-	30	130	1.3	1,200,000	Y
	-	-	4	13	Zr	-	30	130	1.3	1,200,000	Y
	-	-	4	13	Ti	-	30	130	1.3	1,200,000	Y
Gehrke et al. [37] 2006	I	-	4.5	18	Zi	-	100–450	-	15	5,000,000	Y
	I	-	4.5	18	Zi	-	100–450	-	15	5,000,000	Y
	I	-	4.5	18	Zi	-	100–450	-	15	5,000,000	Y
Kohal et al. [38] 2009	M	Zr	-	-	Zi	-	45	-	-	1,200,000	Y
	M	Zr	-	-	Zi	-	45	-	-	1,200,000	Y
	M	Ti	-	-	P	-	45	-	-	1,200,000	Y
Scarano et al. [39] 2010	M	Ti	4	13	-	-	5–230	30	4	1,000,000	N

Table 3. *Cont.*

Author/Year	Implant				Abutment		Applied Load			Cycles	Failure
	Connection	Material	Diameter	Length	Material	Length	Magnitude (N)	Angulation (°)	Frequency (Hz)		
Magne et al. [40] 2011	I	-	-	12	Metal	12	80–280	30	5	20,000	Y
Dittmer et al. [20] 2012	I	-	4.5	13	Ti	1.5	100	30	2	1,000,000	Y
	I	-	4.5	13	Ti	4.1	100	30	2	1,000,000	Y
	E	-	4.3	13	-	11	100	30	2	1,000,000	Y
	I	-	4.5	14	-	1	100	30	2	1,000,000	Y
	E	-	4	13	-	1	100	30	2	1,000,000	Y
	I	-	4.1	14	Ti	5.5	100	30	2	1,000,000	Y
Stimmelmayr et al. [42] 2012	I	Ti	3.8	13	Ti	10	100	-	1.2	1,200,000	N
	I	Ti	3.8	13	Zr	10	100	-	1.2	1,200,000	N
Pintinha et al. [44] 2013	I	Ti	4	10	Ti	8.7	100 ± 5	20	2	500	N
	I	Ti	4	10	Ti	9	100 ± 5	20	2	500	N
Rosentritt et al. [46] 2014	I	Zr	4.1	10	Zr	-	50	45	1.6	1,200,000	Y
	I	Zr	4	10	Zr	-	50	45	1.6	1,200,000	N
	I	Zr	4.1	11	Zr	-	50	45	1.6	1,200,000	Y
	I	Zr	4.1	14	Ti	-	50	45	1.6	1,200,000	Y
	I	Ti	4	15	Ti	-	50	45	1.6	1,200,000	N
	I	Zr	4.1	12	Zr	-	50	45	1.6	1,200,000	N
	I	Zr	4.5	10	Zr	-	50	45	1.6	1,200,000	N
Mitsias et al. [47] 2015	M	-	-	-	Y-TZP	-	400	30	-	100,000	Y
	M	-	-	-	Y-TPZ	-	400	30	-	100,000	Y
Spies et al. [48] 2015	E	ATZ	4.4/4.1/4.2	12	-	6	98	-	2	10,000,000	N
	I	Y-TZP-A	4.1	12	Y-TZP-A	6	98	-	2	10,000,000	N
	E	Y-TZP-A	4.2	12	ATZ	6	98	-	2	10,000,000	N
Guilherme et al. [49] 2016	M	-	4.3	10	Zr	-	150–200	-	2	100	N
	M	-	4.3	10	LD	-	150–200	-	2	100	N
	M	-	4.3	10	R-BC	-	150–200	-	2	100	N
Preis et al. [50] 2016	I	Zr	4.1	10	Zr	-	100	45	1.6	3,600,000	N
	I	Zr	4.1	10	Zr	-	100	45	1.6	3,600,000	Y
	I	Zr	3.8	11	Zr	-	100	45	1.6	3,600,000	Y
	I	Zr	4.6	10	Zr	-	100	45	1.6	3,600,000	Y
	I	Zr	4.1	10	Zr	-	100	45	1.6	3,600,000	Y
	I	Ti	4.1	12	Ti	-	100	45	1.6	3,600,000	N

I: Internal connection; E: external connection; M: morse taper connection; Ti: Titanium; Zr: Zirconia; ATZ: alumina-toughened zirconia; Y-TZP-A: yttrium stabilized tetragonal zirconium dioxide polycrystal doped with alumina; LD: lithium disilicate; R-BC: resin-based composite; Y: Yes; N: No; -: No data.

Table 4. Variables analyzed from the answer to question B.

Author/Year	Implant				Abutment		Applied Load			Cycles	Failure
	Connection	Material	Diameter	Length	Material	Length	Magnitude	Angulation	Frequency		
Balfour et al. [33] 1995	E	Ti	-	-	Ti	-	242	30	14	5,000,000	Y
	I	Ti	-	-	Ti	-	400	30	14	5,000,000	Y
	I	Ti	-	-	Ti	-	367	30	14	5,000,000	Y
Khraisat et al. [34] 2002	E	Ti	4	10	Ti	3	100	90	1,25	1,800,000	Y (1,178,023 and 1,733,526)
	M	Ti	4.1	10	Ti	10	100	90	1,25	1,800,000	Y (more 1,800,000)
Seetoh et al. [41] 2011	M	-	4.5	15	Zr/Ti	-	21	45	10	5,000,000	Y
	I	-	4	15	Zr/Ti	-	21	45	10	5,000,000	Y
	M	-	4.1	14	Zr/Ti	-	21	45	10	5,000,000	Y
Foong et al. [3] 2013	I	Ti	4	9	Ti	1.5	50–400	30	2 to 5	5,000–20,000	Y (mean of 81,935)
	I	Ti	4	9	Zr	1.5	50–400	30	2 to 5	5,000–20,000	Y (mean of 26,926)
Marchetti et al. [45] 2014	I	Ti	3.8	13	Ti	-	400	30 ± 2	15	5,000,000	Y (12,678 and 15,387)
	I	Ti	3.8	13	Ti	-	300	30 ± 2	15	5,000,000	Y (more 27,732)
Mitsias et al. [47] 2015	M	-	-	-	Y-TZP	-	400	30	-	100,000	Y (less than 50,000)
	M	-	-	-	Y-TPZ	-	400	30	-	100,000	Y (less than 50,000)

I: Internal connection; E: external connection; M: morse taper connection; Ti: Titanium; Zr: Zirconia; Y: Yes; N: No; -: No data.

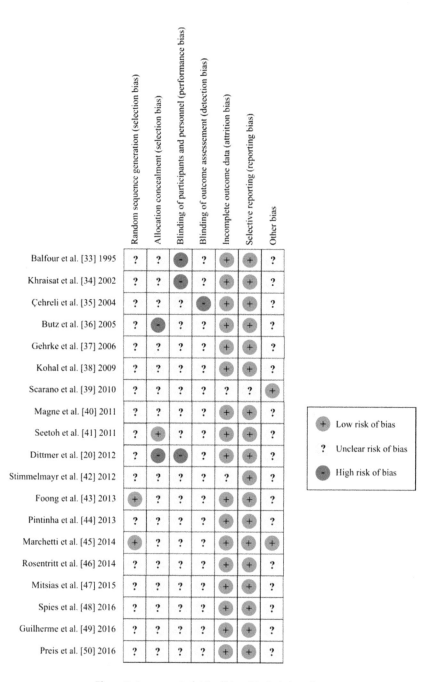

Figure 3. Assessment of risk of bias of included studies.

4. Discussion

There are several factors that influence the behavior of dental implants such as the biological effects of the location and magnitude of applied force [51], occlusal forces following implant treatment [52], immediate or early implant loading [53], the influence of bone quality, effects of prosthesis type, prosthesis material, or implant support [51].

The clinical long-term success of restorations in oral implantology depends, among other factors on the stable connection between the dental implant and the abutment [54]. Therefore, a good knowledge of the biomechanical behavior of dental implants is essential for clinical decision making and thus, avoid mechanical failures. These are mainly due to fatigue caused by overload or loss of bone around the implant [55–57].

The most common connections between the implant and the prosthesis are the external, internal hexagon and morse taper connections [58]. The main advantages of the external connection are the compatibility with a wide variety of implants, its economical price, the long-term follow-up data available, and the literature provides solutions to the main drawbacks associated with its use. However, among the main disadvantages of this connection are the loosening of the screws, the possibility of screw fracture, worse aesthetic results, and an inadequate microbial seal [59–61].

The appearance of the internal connections was due to the interest in trying to reduce the aesthetic and microbial filtration problems of the external connection, while improving the behavior of the implant and peri-implant bone against masticatory forces [60]. The main advantages of these internal connections are less or no screw loosening, less risk of screw fracture, aesthetic improvement, microbial sealing, and the stability of the implant-prosthesis connection. Among the disadvantages are a higher economic cost and long-term monitoring data lower than the external connection [59,62].

In the morse taper connections, all the components that make up the implantoprosthesis assembly behave as a single whole, the forces are adequately distributed, the stability of the prosthesis is guaranteed, and the areas of greatest mechanical suffering are protected, such as the crestal region of the implant [17,63].

Other factors that influence the fracture of the dental implant are the materials, the diameter and length of the implant, the material and length of the abutment, the applied force, its frequency, and the angulation with respect to the implant [51]. This work includes all the parameters described above even though there is a considerable amount of heterogeneity.

Titanium is still the most used material and it is also supported by long-term clinical studies [4], but also in vitro studies since 1995 [33] introduced zirconium with the gold standard, titanium, to evaluate the fatigue of dental implants. Although the results were favorable, its biggest drawback is its high degree of quality requirement during production and its delicate clinical management [4].

The materials most used for dental implants were titanium and zirconia, with ranges of measures in relation to diameter and height, of 3.8 mm to 4.6 mm and 10 mm to 15 mm, respectively. For the abutment the most used materials were also titanium and zirconia but in the one study [49] also used lithium disilicate and resin-based composite. There is no homogeneous criterion regarding the magnitude, angulation, frequency, and number of cycles applied in the dynamic loads to the implants. Therefore, with the data reported it is not possible to report conclusive results from the different in vitro studies analyzed.

There are several engineering methods to evaluate the fatigue behavior of dental implants such as finite elements [64], mathematical models for probabilistic fatigue [65], or other in vitro studies. These type of study designs have developed, before their use in patients, new materials to understand their physical, chemical, mechanical, and biological properties. These designs can be developed under normal conditions or under other conditions such as in water [66,67] or hydrothermal [68,69] environments.

There is not a large number of studies evaluating fatigue in dental implants in the scientific literature. However, a greater number of in vitro studies which evaluated the same objectives under normal conditions have been addressed. It is therefore desirable to conduct a systematic review.

The main disadvantage of the evaluation of the articles of this work is the high heterogeneity of the confounding factors collected as the different materials, lengths, diameters, or dynamic loading conditions. There is no homogeneity in the design criteria established by each study in this regard.

With the objective of obtaining more clinically relevant information, future studies should incorporate and analyze the same parameters. Otherwise, there heterogeneity will continue creating doubt in the scientific literature in the field of dentistry [2]. Therefore and, in view of the results obtained in this systematic review, future in vitro analysis should have the same implants dimensions (diameter and length) and cyclic loading conditions (number of cycles, magnitude of force, angle, and medium). The no homogeneity found in these studies contribute to realize that it is necessary to standardize the criteria for carrying out the studies in order to make more concise and reliable comparisons.

Some studies [39,45,46,48,49] use ISO 14801 [24] which specifies the conditions that each type of implant must support to obtain certification. However, these conditions indicate that they should exceed a minimum number of cycles but say nothing about how to conduct experimental studies of implants that are already on the market, and therefore have already been certified.

In systematic reviews, a qualitative analysis of the included studies is required. This is done through the risk assessment of bias. This ensures that the data collected and analyzed have been managed in a controlled manner, avoiding all possible methodological errors in clinical trials. When the data are homogeneous, in addition, a quantitative analysis can be carried out, through meta-analysis [31].

This means that each study conditions to decide which subjects its implants, without any specific criteria, which makes it very difficult to know which of the implants have better mechanical behavior under certain conditions. This fact is evident in Tables 3 and 4 in which it is possible to observe the great variety of conditions used in the articles included in this review.

We used the Cochrane Handbook tool to assess the risk of bias in the studies, noting that in most domains, no data are given that give transparency to the studies. Generally, the criteria for randomization and allocation masking, and blinding of staff and data assessors are not indicated. Together with these detected defects and under recommendations some authors [70] in vitro studies should be treated to promote the quality of the tests: simple size calculation, meaningful difference between groups, sample preparation and handling, allocation sequence, randomization and blinding, statistical analysis.

The heterogeneity of data available in the scientific literature does not allow a meta-analysis in the field of in vitro fatigue and fracture of dental implants. As in the design of clinical trials in humans with the CONSORT (Consolidated Standards of Reporting Trials) guidelines [71], we consider it advisable to follow guidelines for in vitro studies such as the CRIS (Checklist for Reporting in vitro Studies) guidelines [70]. Also, define the criteria and conditions of applied loads (magnitude, angle, frequency, cycles ...) and are contained in an ISO standard.

Nevertheless, we have found, in the present review that the internal connections [42,44], and those based on the morse taper system [34,39,49,72] show a better performance against resistance to fracture in the dental implant compared to the external connection [34,48]. Moreover, the results revealed that the implant and the abutment have better behavior if both materials are the same.

In addition, these studies assessed a range of materials, but the most frequently used materials are still in order of use, titanium and zirconium, with a behavior similar to fatigue.

5. Conclusions

The limitations found in this review do not allow us to report consistent evidence. The results suggest that the internal and morse connections are the best for resisting the fracture of the dental implants and the most commonly used materials are titanium and zirconium. However, it is necessary to unify criteria in the methodological design of this type of in vitro studies.

Author Contributions: R.R. developed the main part of the review, created the search strategy, selected the articles, performed the statistical analysis and wrote part of the paper. M.P.-P. selected the articles, evaluated the methodological quality of the studies and wrote part of the paper. A.J.R. evaluated the methodological quality of the studies. J.C.P.-F. provided critical analysis and interpretation of data.

Acknowledgments: The authors received financial support of grants A-274 (Instradent Iberia® S.A., Alcobendas, Spain) and A-285 (Proclinic® S.A, Zaragoza, Spain).

Conflicts of Interest: The authors declare no conflict of interest.

References

1. Tian, K.; Chen, J.; Han, L.; Yang, J.; Huang, W.; Wu, D. Angled abutments result in increased or decreased stress on surrounding bone of single-unit dental implants: A finite element analysis. *Med. Eng. Phys.* **2012**, *34*, 1526–1531. [CrossRef] [PubMed]
2. Coray, R.; Zeltner, M.; Özcan, M. Fracture strength of implant abutments after fatigue testing: A systematic review and a meta-analysis. *J. Mech. Behav. Biomed. Mater.* **2016**, *62*, 333–346. [CrossRef] [PubMed]
3. Elias, C.N.; Fernandes, D.J.; Resende, C.R.S.; Roestel, J. Mechanical properties, surface morphology and stability of a modified commercially pure high strength titanium alloy for dental implants. *Dent. Mater.* **2015**, *31*, e1–e13. [CrossRef] [PubMed]
4. Osman, R.B.; Swain, M.V. A Critical Review of Dental Implant Materials with an Emphasis on Titanium versus Zirconia. *Materials* **2015**, *8*, 932–958. [CrossRef] [PubMed]
5. Ottria, L.; Lauritano, D.; Andreasi Bassi, M.; Palmieri, A.; Candotto, V.; Tagliabue, A.; Tettamanti, L. Mechanical, chemical and biological aspects of titanium and titanium alloys in implant dentistry. *J. Biol. Regul. Homeost. Agents* **2018**, *32*, 81–90. [PubMed]
6. Kirmanidou, Y.; Sidira, M.; Drosou, M.-E.; Bennani, V.; Bakopoulou, A.; Tsouknidas, A.; Michailidis, N.; Michalakis, K. New Ti-Alloys and Surface Modifications to Improve the Mechanical Properties and the Biological Response to Orthopedic and Dental Implants: A Review. *Biomed Res. Int.* **2016**, *2016*, 2908570. [CrossRef] [PubMed]
7. Kent, D.; Wang, G.; Dargusch, M. Effects of phase stability and processing on the mechanical properties of Ti-Nb based beta Ti alloys. *J. Mech. Behav. Biomed. Mater.* **2013**, *28*, 15–25. [CrossRef] [PubMed]
8. Hatamleh, M.M.; Wu, X.; Alnazzawi, A.; Watson, J.; Watts, D. Surface characteristics and biocompatibility of cranioplasty titanium implants following different surface treatments. *Dent. Mater.* **2018**, *34*, 676–683. [CrossRef] [PubMed]
9. Wennerberg, A. The importance of surface roughness for implant incorporation. *Int. J. Mach. Tools Manuf.* **1998**, *38*, 657–662. [CrossRef]
10. Komasa, S.; Taguchi, Y.; Nishida, H.; Tanaka, M.; Kawazoe, T. Bioactivity of nanostructure on titanium surface modified by chemical processing at room temperature. *J. Prosthodont. Res.* **2012**, *56*, 170–177. [CrossRef] [PubMed]
11. Kubasiewicz-Ross, P.; Dominiak, M.; Gedrange, T.; Botzenhart, U.U. Zirconium: The material of the future in modern implantology. *Adv. Clin. Exp. Med.* **2017**, *26*, 533–537. [CrossRef] [PubMed]
12. Freitas, G.; Hirata, R.; Bonfante, E.; Tovar, N.; Coelho, P. Survival Probability of Narrow and Standard-Diameter Implants with Different Implant-Abutment Connection Designs. *Int. J. Prosthodont.* **2016**, *29*, 179–185. [CrossRef] [PubMed]
13. Bartlett, D. Implants for life? A critical review of implant-supported restorations. *J. Dent.* **2007**, *35*, 768–772. [CrossRef] [PubMed]
14. Pardal-Pelaez, B.; Montero, J. Preload loss of abutment screws after dynamic fatigue in single implant-supported restorations. A systematic review. *J. Clin. Exp. Dent.* **2017**, *9*, e1355–e1361. [CrossRef] [PubMed]
15. Gehrke, S.A.; Delgado-Ruiz, R.A.; Prados Frutos, J.C.; Prados-Privado, M.; Dedavid, B.A.; Granero Marin, J.M.; Calvo Guirado, J.L. Misfit of Three Different Implant-Abutment Connections Before and After Cyclic Load Application: An In Vitro Study. *Int. J. Oral Maxillofac. Implant.* **2017**, *32*, 822–829. [CrossRef] [PubMed]
16. Prados-Privado, M.; Bea, J.A.; Rojo, R.; Gehrke, S.A.; Calvo-Guirado, J.L.; Prados-Frutos, J.C. A New Model to Study Fatigue in Dental Implants Based on Probabilistic Finite Elements and Cumulative Damage Model. *Appl. Bionics Biomech.* **2017**, *2017*. [CrossRef] [PubMed]
17. Schmitt, C.M.; Nogueira-Filho, G.; Tenenbaum, H.C.; Lai, J.Y.; Brito, C.; Döring, H.; Nonhoff, J. Performance of conical abutment (Morse Taper) connection implants: A systematic review. *J. Biomed. Mater. Res. Part A* **2014**, *102*, 552–574. [CrossRef] [PubMed]

18. Sailer, I.; Sailer, T.; Stawarczyk, B.; Jung, R.E.; Hämmerle, C.H.F. In vitro study of the influence of the type of connection on the fracture load of zirconia abutments with internal and external implant-abutment connections. *Int. J. Oral Maxillofac. Implant.* **2009**, *24*, 850–858. [CrossRef]

19. Kitagawa, T.; Tanimoto, Y.; Odaki, M.; Nemoto, K.; Aida, M. Influence of implant/abutment joint designs on abutment screw loosening in a dental implant system. *J. Biomed. Mater. Res. Part B Appl. Biomater.* **2005**, *75B*, 457–463. [CrossRef] [PubMed]

20. Dittmer, M.P.; Dittmer, S.; Borchers, L.; Kohorst, P.; Stiesch, M. Influence of the interface design on the yield force of the implant–abutment complex before and after cyclic mechanical loading. *J. Prosthodont. Res.* **2012**, *56*, 19–24. [CrossRef] [PubMed]

21. Ritter, J.E. Critique of test methods for lifetime predictions. *Dent. Mater.* **1995**, *11*, 147–151. [CrossRef]

22. Marx, R.; Jungwirth, F.; Walter, P.-O. Threshold intensity factors as lower boundaries for crack propagation in ceramics. *Biomed. Eng. Online* **2004**, *3*, 41. [CrossRef] [PubMed]

23. Alqahtani, F.; Flinton, R. Postfatigue fracture resistance of modified prefabricated zirconia implant abutments. *J. Prosthet. Dent.* **2014**, *112*, 299–305. [CrossRef] [PubMed]

24. Organization, I.S. *ISO 14801: Dentistry—Implants—Dynamic Fatigue Test for Endosseous Dental Implants*; ISO: Geneve, Switzerland, 2007.

25. Marchetti, E.; Ratta, S.; Mummolo, S.; Tecco, S.; Pecci, R.; Bedini, R.; Marzo, G. Mechanical Reliability Evaluation of an Oral Implant-Abutment System According to UNI EN ISO 14801 Fatigue Test Protocol. *Implant Dent.* **2016**, *25*, 613–618. [CrossRef] [PubMed]

26. Lee, C.K.; Karl, M.; Kelly, J.R. Evaluation of test protocol variables for dental implant fatigue research. *Dent. Mater.* **2009**, *25*, 1419–1425. [CrossRef] [PubMed]

27. Snauwaert, K.; Duyck, J.; van Steenberghe, D.; Quirynen, M.; Naert, I. Time dependent failure rate and marginal bone loss of implant supported prostheses: A 15-year follow-up study. *Clin. Oral Investig.* **2000**, *4*, 0013–0020. [CrossRef]

28. Hasan, I.; Bourauel, C.; Mundt, T.; Stark, H.; Heinemann, F. Biomechanics and load resistance of small-diameter and mini dental implants: A review of literature. *Biomed. Tech. Eng.* **2014**, *59*, 1–5. [CrossRef] [PubMed]

29. Moher, D.; Altman, D.G.; Liberati, A.; Tetzlaff, J. PRISMA statement. *Epidemiology* **2011**, *22*, 128. [CrossRef] [PubMed]

30. Centre for Rewies and Dissemination, University of York. *Systematic Reviews: CRD Guidance for Undertaking Reviews in Health Care*; University of York: York, UK, 2009.

31. Higgins, J.P.T.; Altman, D.G. Assessing risk of bias in included studies. In *Cochrane Handbook for Systematic Reviews of Interventions*; Higgins, J.P.T., Green, S., Eds.; Wiley: Hoboken, NJ, USA, 2008; pp. 187–241.

32. Landis, J.R.; Koch, G.G. The measurement of observer agreement for categorical data. *Biometrics* **1977**, *33*, 159–174. [CrossRef] [PubMed]

33. Balfour, A.; O'Brien, G.R. Comparative study of antirotational single tooth abutments. *J. Prosthet. Dent.* **1995**, *73*, 36–43. [CrossRef]

34. Khraisat, A.; Stegaroiu, R.; Nomura, S.; Miyakawa, O. Fatigue resistance of two implant/abutment joint designs. *J. Prosthet. Dent.* **2002**, *88*, 604–610. [CrossRef] [PubMed]

35. Cehreli, M.C.; Akca, K.; Iplikcioglu, H.; Sahin, S. Dynamic fatigue resistance of implant-abutment junction in an internally notched morse-taper oral implant: Influence of abutment design. *Clin. Oral Implant. Res.* **2004**, *15*, 459–465. [CrossRef] [PubMed]

36. Butz, F.; Heydecke, G.; Okutan, M.; Strub, J.R. Survival rate, fracture strength and failure mode of ceramic implant abutments after chewing simulation. *J. Oral Rehabil.* **2005**, *32*, 838–843. [CrossRef] [PubMed]

37. Gehrke, P.; Dhom, G.; Brunner, J.; Wolf, D.; Degidi, M.; Piattelli, A. Zirconium implant abutments: Fracture strength and influence of cyclic loading on retaining-screw loosening. *Quintessence Int.* **2006**, *37*, 19–26. [PubMed]

38. Kohal, R.-J.; Finke, H.C.; Klaus, G. Stability of prototype two-piece zirconia and titanium implants after artificial aging: An in vitro pilot study. *Clin. Implant Dent. Relat. Res.* **2009**, *11*, 323–329. [CrossRef] [PubMed]

39. Scarano, A.; Sacco, M.L.; Di Iorio, D.; Amoruso, M.; Mancino, C. Valutazione della resistenza a fatica ciclica di una connessione impianto-abutment cone morse e avvitata. *Ital. Oral Surg.* **2010**, *9*, 173–179. [CrossRef]

40. Magne, P.; Oderich, E.; Boff, L.L.; Cardoso, A.C.; Belser, U.C. Fatigue resistance and failure mode of CAD/CAM composite resin implant abutments restored with type III composite resin and porcelain veneers. *Clin. Oral Implant. Res.* **2011**, *22*, 1275–1281. [CrossRef] [PubMed]

41. Seetoh, Y.L.; Tan, K.B.; Chua, E.K.; Quek, H.C.; Nicholls, J.I. Load fatigue performance of conical implant-abutment connections. *Int. J. Oral Maxillofac. Implant.* **2011**, *26*, 797–806.

42. Stimmelmayr, M.; Edelhoff, D.; Güth, J.-F.; Erdelt, K.; Happe, A.; Beuer, F. Wear at the titanium–titanium and the titanium–zirconia implant–abutment interface: A comparative in vitro study. *Dent. Mater.* **2012**, *28*, 1215–1220. [CrossRef] [PubMed]

43. Foong, J.K.W.; Judge, R.B.; Palamara, J.E.; Swain, M.V. Fracture resistance of titanium and zirconia abutments: An in vitro study. *J. Prosthet. Dent.* **2013**, *109*, 304–312. [CrossRef]

44. Pintinha, M.; Camarini, E.T.; Sábio, S.; Pereira, J.R. Effect of mechanical loading on the removal torque of different types of tapered connection abutments for dental implants. *J. Prosthet. Dent.* **2013**, *110*, 383–388. [CrossRef] [PubMed]

45. Marchetti, E.; Ratta, S.; Mummolo, S.; Tecco, S.; Pecci, R.; Bedini, R.; Marzo, G. Evaluation of an Endosseous Oral Implant System According to UNI EN ISO 14801 Fatigue Test Protocol. *Implant Dent.* **2014**, *23*, 665–671. [CrossRef] [PubMed]

46. Rosentritt, M.; Hagemann, A.; Hahnel, S.; Behr, M.; Preis, V. In vitro performance of zirconia and titanium implant/abutment systems for anterior application. *J. Dent.* **2014**, *42*, 1019–1026. [CrossRef] [PubMed]

47. Mitsias, M.E.; Thompson, V.P.; Pines, M.; Silva, N.R.F.A. Reliability and failure modes of two Y-TZP abutment designs. *Int. J. Prosthodont.* **2015**, *28*, 75–78. [CrossRef] [PubMed]

48. Spies, B.C.; Nold, J.; Vach, K.; Kohal, R.-J. Two-piece zirconia oral implants withstand masticatory loads: An investigation in the artificial mouth. *J. Mech. Behav. Biomed. Mater.* **2016**, *53*, 1–10. [CrossRef] [PubMed]

49. Guilherme, N.M.; Chung, K.-H.; Flinn, B.D.; Zheng, C.; Raigrodski, A.J. Assessment of reliability of CAD-CAM tooth-colored implant custom abutments. *J. Prosthet. Dent.* **2016**, *116*, 206–213. [CrossRef] [PubMed]

50. Preis, V.; Kammermeier, A.; Handel, G.; Rosentritt, M. In vitro performance of two-piece zirconia implant systems for anterior application. *Dent. Mater.* **2016**, *32*, 765–774. [CrossRef] [PubMed]

51. Sahin, S.; Cehreli, M.C.; Yalcin, E. The influence of functional forces on the biomechanics of implant-supported prostheses—A review. *J. Dent.* **2002**, *30*, 271–282. [CrossRef]

52. Flanagan, D. Bite force and dental implant treatment: A short review. *Med. Devices* **2017**, *10*, 141–148. [CrossRef] [PubMed]

53. Chrcanovic, B.R.; Kisch, J.; Albrektsson, T.; Wennerberg, A. Factors Influencing Early Dental Implant Failures. *J. Dent. Res.* **2016**, *95*, 995–1002. [CrossRef] [PubMed]

54. Hoyer, S.A.; Stanford, C.M.; Buranadham, S.; Fridrich, T.; Wagner, J.; Gratton, D. Dynamic fatigue properties of the dental implant-abutment interface: Joint opening in wide-diameter versus standard-diameter hex-type implants. *J. Prosthet. Dent.* **2001**, *85*, 599–607. [CrossRef] [PubMed]

55. Piattelli, A.; Scarano, A.; Piattelli, M.; Vaia, E.; Matarasso, S. Hollow implants retrieved for fracture: A light and scanning electron microscope analysis of 4 cases. *J. Periodontol.* **1998**, *69*, 185–189. [CrossRef] [PubMed]

56. Tolman, D.E.; Laney, W.R. Tissue-integrated prosthesis complications. *Int. J. Oral Maxillofac. Implants* **1992**, *7*, 477–484. [CrossRef] [PubMed]

57. Steinebrunner, L.; Wolfart, S.; Ludwig, K.; Kern, M. Implant-abutment interface design affects fatigue and fracture strength of implants. *Clin. Oral Implant. Res.* **2008**, *19*, 1276–1284. [CrossRef] [PubMed]

58. Pita, M.S.; Anchieta, R.B.; Barao, V.A.; Garcia, I.R., Jr.; Pedrazzi, V.; Assuncao, W.G. Prosthetic platforms in implant dentistry. *J. Craniofac. Surg.* **2011**, *22*, 2327–2331. [CrossRef] [PubMed]

59. Gaviria, L.; Salcido, J.P.; Guda, T.; Ong, J.L. Current trends in dental implants. *J. Korean Assoc. Oral Maxillofac. Surg.* **2014**, *40*, 50–60. [CrossRef] [PubMed]

60. Finger, I.M.; Castellon, P.; Block, M.; Elian, N. The evolution of external and internal implant/abutment connections. *Pract. Proced. Aesthet. Dent.* **2003**, *15*, 625–632. [PubMed]

61. Binon, P.P. Implants and components: Entering the new millennium. *Int. J. Oral Maxillofac. Implant.* **2000**, *15*, 76–94.

62. Gracis, S.; Michalakis, K.; Vigolo, P.; Vult von Steyern, P.; Zwahlen, M.; Sailer, I. Internal vs. external connections for abutments/reconstructions: A systematic review. *Clin. Oral Implant. Res.* **2012**, *23* (Suppl. 6), 202–216. [CrossRef] [PubMed]

63. Macedo, J.P.; Pereira, J.; Vahey, B.R.; Henriques, B.; Benfatti, C.A.; Magini, R.S.; Lopez-Lopez, J.; Souza, J.C. Morse taper dental implants and platform switching: The new paradigm in oral implantology. *Eur. J. Dent.* **2016**, *10*, 148–154. [CrossRef] [PubMed]

64. Cheng, Y.C.; Lin, D.H.; Jiang, C.P.; Lee, S.Y. Design improvement and dynamic finite element analysis of novel ITI dental implant under dynamic chewing loads. *Biomed. Mater. Eng.* **2015**, *26* (Suppl. 1), S555–S561. [CrossRef] [PubMed]

65. Prados-Privado, M.; Prados-Frutos, J.C.; Calvo-Guirado, J.L.; Bea, J.A. A random fatigue of mechanize titanium abutment studied with Markoff chain and stochastic finite element formulation. *Comput. Methods Biomech. Biomed. Eng.* **2016**, *19*, 1583–1591. [CrossRef] [PubMed]

66. Anchieta, R.B.; Machado, L.S.; Hirata, R.; Bonfante, E.A.; Coelho, P.G. Platform-Switching for Cemented Versus Screwed Fixed Dental Prostheses: Reliability and Failure Modes: An In Vitro Study. *Clin. Implant Dent. Relat. Res.* **2016**, *18*, 830–839. [CrossRef] [PubMed]

67. Bordin, D.; Bergamo, E.T.P.; Fardin, V.P.; Coelho, P.G.; Bonfante, E.A. Fracture strength and probability of survival of narrow and extra-narrow dental implants after fatigue testing: In vitro and in silico analysis. *J. Mech. Behav. Biomed. Mater.* **2017**, *71*, 244–249. [CrossRef] [PubMed]

68. Kim, J.-W.; Covel, N.S.; Guess, P.C.; Rekow, E.D.; Zhang, Y. Concerns of hydrothermal degradation in CAD/CAM zirconia. *J. Dent. Res.* **2010**, *89*, 91–95. [CrossRef] [PubMed]

69. Kawai, Y.; Uo, M.; Wang, Y.; Kono, S.; Ohnuki, S.; Watari, F. Phase transformation of zirconia ceramics by hydrothermal degradation. *Dent. Mater. J.* **2011**, *30*, 286–292. [CrossRef] [PubMed]

70. Krithikadatta, J.; Gopikrishna, V.; Datta, M. CRIS Guidelines (Checklist for Reporting In-vitro Studies): A concept note on the need for standardized guidelines for improving quality and transparency in reporting in-vitro studies in experimental dental research. *J. Conserv. Dent.* **2014**, *17*, 301–304. [CrossRef] [PubMed]

71. Palmas, W. The CONSORT guidelines for noninferiority trials should be updated to go beyond the absolute risk difference. *J. Clin. Epidemiol.* **2017**, *83*, 6–7. [CrossRef] [PubMed]

72. Machado, L.S.; Bonfante, E.A.; Anchieta, R.B.; Yamaguchi, S.; Coelho, P.G. Implant-abutment connection designs for anterior crowns: Reliability and failure modes. *Implant Dent.* **2013**, *22*, 540–545. [CrossRef] [PubMed]

Article

Insights into Machining of a β Titanium Biomedical Alloy from Chip Microstructures

Damon Kent [1,2,3,*], **Rizwan Rahman Rashid** [4,5], **Michael Bermingham** [2,3], **Hooyar Attar** [2], **Shoujin Sun** [4] and **Matthew Dargusch** [2,3]

[1]   School of Science and Engineering, University of the Sunshine Coast, Maroochydore DC 4558, Australia
[2]   Queensland Centre for Advanced Materials Processing and Manufacturing (AMPAM),
     The University of Queensland, St. Lucia 4072, Australia; m.bermingham@uq.edu.au (M.B.);
     h.attar@uq.edu.au (H.A.); m.dargusch@uq.edu.au (M.D.)
[3]   ARC Research Hub for Advanced Manufacturing of Medical Devices, St. Lucia 4072, Australia
[4]   School of Engineering, Faculty of Science, Engineering and Technology, Swinburne University of Technology,
     Victoria 3122, Australia; rrahmanrashid@swin.edu.au (R.R.R.); ssun@swin.edu.au (S.S.)
[5]   Defence Materials Technology Centre, Victoria 3122, Australia
*   Correspondence: dkent@usc.edu.au; Tel.: +61-5456-5267

Received: 17 August 2018; Accepted: 10 September 2018; Published: 11 September 2018

Abstract: New metastable β titanium alloys are receiving increasing attention due to their excellent biomechanical properties and machinability is critical to their uptake. In this study, machining chip microstructure has been investigated to gain an understanding of strain and temperature fields during cutting. For higher cutting speeds, ≥60 m/min, the chips have segmented morphologies characterised by a serrated appearance. High levels of strain in the primary shear zone promote formation of expanded shear band regions between segments which exhibit intensive refinement of the β phase down to grain sizes below 100 nm. The presence of both α and β phases across the expanded shear band suggests that temperatures during cutting are in the range of 400–600 °C. For the secondary shear zone, very large strains at the cutting interface result in heavily refined and approximately equiaxed nanocrystalline β grains with sizes around 20–50 nm, while further from the interface the β grains become highly elongated in the shear direction. An absence of the α phase in the region immediately adjacent to the cutting interface indicates recrystallization during cutting and temperatures in excess of the 720 °C β transus temperature.

Keywords: machining; titanium; temperature; strain; grain refinement; ultrafine; nanocrystalline

1. Introduction

β titanium alloys possess high strength to weight, excellent toughness, corrosion resistance and biocompatibility and so have excellent potential for a wide range of biomedical applications [1]. In the last decade, there has been significant focus on the development of a variety of new metastable β titanium alloys with lower Young's moduli approaching that of human bone. These alloys employ various combinations of elements to stabilise the body-centred cubic β titanium phase and can exhibit both shape memory and pseudoelastic behaviours [2,3]. A metastable Ti-Nb based β titanium alloy (Ti-25Nb-3Mo-3Zr-2Sn wt.%) with excellent mechanical and biological compatibility has recently been the subject of extensive research and development by the authors [4–6].

Biomedical components manufactured from titanium alloys typically require machining to achieve their required form, size and surface finish. However, machining can be problematic, particularly at high speeds, due to issues with build-up of heat at the cutting zone associated with titanium's relatively low thermal conductivity and high levels of chemical affinity which lead to reaction with and 'sticking' to the cutting tool materials [7,8]. Typically, more than 70% of the heat generated during machining is delivered to the cutting tool, intensifying the degree of chemical interaction between the tool and workpiece [9,10]. For these reasons, most titanium alloys are considered difficult to machine and much of the fabrication cost for geometrically complex components may be due to machining. Hence, there is a strong incentive to better understand the machining process to improve material removal for these alloys. A further driver to study these processes comes from observations that the plastic deformation which takes place at the cutting interface can also significantly influence cell viability and adhesion on metallic implant materials [11].

Previously, Rashid et al. studied the machinability of the Ti-25Nb-3Mo-3Zr-2Sn alloy including cutting forces, temperatures and macroscopic chip characteristics [12,13]. For the solution treated and aged Ti-25Nb-3Mo-3Zr-2Sn alloy, the main cutting force decreases from around 600 N at low cutting speeds to around 430 N for speeds above 30 m/min, remaining constant at this level for speeds up to almost 200 m/min. Measurements of the external chip surface temperatures in the cutting zone using infrared thermography revealed that the temperatures are $\leq$300 °C for low surface cutting speeds (below 10 m/min), increasing markedly to more than 700 °C for high surface cutting speeds approaching 200 m/min [12]. The machining chips transition from a continuous form at low cutting speeds to a segmented saw-tooth morphology for surface cutting speeds $\geq$60 m/min [13]. The frequency of shear regions between individual sawtooth segments is associated with significant fluctuations in component forces during machining which exacerbate tool wear.

The shear regions between the sawtooth segments are subject to localised, high strain rate, severe plastic deformation at elevated temperatures. Due to the relatively small chip mass, the metal is effectively quenched as it leaves the cutting zone, preserving the as-machined microstructures. The extreme deformation conditions may result in the formation of nano-crystalline and/or ultrafine-grain microstructures of interest from the perspective of improving fundamental knowledge of severe plastic deformation processes and their associated microstructures, as well as understanding the cutting process. Schneider et al. used focussed ion beam (FIB) specimens from machining chips in conjunction with transmission electron microscopy (TEM) to study the fine microstructural features within the secondary deformation zone from cutting of the $\alpha + \beta$, Ti-6Al-4V (wt.%) alloy [14]. A layered microstructure with fine grains near the cutting interface transitioning to coarse grains toward the free surface was observed. A 10 nm thick recrystallised layer was present at the cutting interface which adjoined a 20 nm thick amorphous layer. To the best of the author's knowledge, similar high level characterisation of the fine scale deformation features formed during machining of the increasingly important β titanium class of alloys has not yet been undertaken.

Recently, the deformation behaviours of the Ti-25Nb-3Mo-3Zr-2Sn (wt.%) alloy under high strain rates ($\approx$1000 s^{-1}), in the order of those encountered in machining, were studied using Split Hopkinson Pressure Bar testing [15]. High strain rates alone did not significantly alter the deformation mechanisms from those occurring under quasistatic strain conditions which involve twinning ({332} <113> and {112} <111> twinning systems) as well as stress induced formation of the α'' and ω phases. The strain hardening behaviour of the alloy was also strain rate insensitive under these conditions due to limited adiabatic heating. However, at elevated deformation temperatures ($\geq$300 °C), the preferred deformation mechanism shifts to dislocation slip due to an increased relative stability of the β phase promoting textural changes in the β grain orientation to those favouring slip, i.e., the <001> and <111> fibre textures [16]. At elevated temperatures the yield stress also significantly reduces due to the cessation of mechanical twinning in association with significant thermal softening. These observations can inform the interpretation of the deformation processes taking place in the chips during machining of the Ti-25Nb-3Mo-3Zr-2Sn alloy. Hence, the aim of this study is to investigate deformation microstructures preserved in the machining chips and to relate these to the strain and temperature fields present during machining. This will assist to better predict, model and optimize machining operations involving β titanium biomedical alloys.

2. Materials and Methods

The investigated alloy has a nominal alloy composition of Ti-25Nb-3Mo-3Zr-2Sn (wt.%). A 25 kg ingot was produced by alloying commercially pure Ti sponge (99.5 wt.% purity), pure Zr bars (99.7 wt.%), pure Sn bars (99.9 wt.%), pure Mo powder (99.8 wt.%) and an intermediate Nb-47 wt.% Ti alloy. The alloy was melted twice by non-consumable arc melting to ensure chemical homogeneity and low levels of impurities. The ingots were forged and then hot rolled to produce cylindrical bars 33 mm in diameter. The bars were solution treated at 750 °C followed by air cooling and ageing at 450 °C for 2 h followed by air cooling.

The machining operation was performed on 3.5 hp Hafco Metal Master lathe (Brisbane, QLD, Australia), Model AL540. A carbide tool CNMX1204A2-SMH13A provided by Sandvik with +15° rake angle, −6° inclination angle and entry angle of 45° was used to machine the Ti-25Nb-3Mo-3Zr-2Sn alloy under dry machining conditions. The machining operation took place under a constant feed rate of 0.19 mm/rev and a constant depth of cut of 1 mm. Microstructural examination was performed on machining chips from cutting with surface cutting speeds around 90 m/min.

The machining chips were mounted and polished with the width direction of the chip perpendicular to the polished surface to reveal the serrated chip cross-section. The chips were etched with Kroll's reagent for observation with scanning electron microscopy (SEM) performed on a JEOL 6460 instrument (Sydney, Australia) equipped with backscatter detector. Hardness testing was conducted on polished specimens using a Struers Vickers microhardness tester (Brisbane, Australia). X-ray diffraction (XRD) was conducted using a Bruker D8 Advance X-ray Diffractometer (Melbourne, Australia) operated at 40 KV and 30 mA, equipped with a graphite monochromator, a Ni-filtered Cu Kα (λ = 1.5406 nm) source and a scintillation counter. Specimens for transmission electron microscopy (TEM) were prepared from transverse mounted sections of the machining chips by dual focused ion beam (FIB) milling using a Zeiss Auriga FIB-SEM (Adelaide, Australia). Sections approximately 100 nm in thickness were milled using a Ga+ beam with typical dimensions of 5 μm × 12 μm. The sections were attached to a C-section copper grid. The TEM was performed using a Philips Tecnai 20 FEG instrument (Brisbane, Australia).

3. Results

3.1. Workpiece Material

The solution treated and aged Ti-25Nb-3Mo-3Zr-2Sn alloy shown in Figure 1a with XRD phase analysis in b consists of β grains with grain sizes in the order of 50 μm and lath shaped α precipitates (the dark phase in the SEM image) located primarily around the grain boundaries and protruding into the β grains. Some α laths are also present within the interior of the β grains. The solution treated and aged alloy has a hardness of 265 ± 5 HV, an ultimate tensile strength of approximately 800 MPa with typical tensile elongation of around 8% [12].

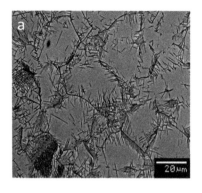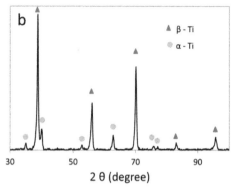

Figure 1. Scanning electron microscopy (SEM) image (**a**) and X-ray diffraction (XRD) spectrum (**b**) from the solution treated and aged workpiece material.

3.2. Machining Chip Characteristics

Cross-sections of Ti-25Nb-3Mo-3Zr-2Sn chips produced for cutting surface speeds of approximately 90 m/min are shown in Figure 2. Within this cutting regime the chips have segmented morphologies characterised by a serrated appearance with bands of severe plastic deformation, referred to herewith as expanded shear band regions, with more limited deformation in adjoining regions. According to previous research by the authors', at these cutting speeds the undeformed chip surface length, i.e., the distance between the regions of severe deformation measured from the top surface of the chip, are approximately 0.08 mm while the average chip thickness is approximately 0.15 mm and the chip roughness ratio is around 0.2 [12,13].

The expanded shear band regions between the sawtooth chips feature extensive deformation as indicated in Figure 2a, while the regions outside these bands are subject to more limited deformation and the original β grain structure of the workpiece material remains discernable, examples of which are indicated in Figure 2b. Typical microhardness values measured from within the expanded shear band regions were 398 ± 15 HV, while adjoining regions with more limited deformation had hardness of around 292 ± 8 HV which is still substantially harder than the initial starting material (265 ± 5 HV). As is also the case for formation of continuous chips, during formation of segmented chips deformation of the material occurs ahead of the tool in the region referred to as the primary shear zone. The transition from continuous to segmented chip morphologies, featuring thermoplastic instability-induced shear banding, emerges once the smooth chip flow becomes insufficient to dissipate the energy through homogeneous plastic flow [17]. The other significant area of deformation in the chip is the secondary shear zone shown in Figure 2c, which is the region adjoining the tool rake face during cutting. Hardness measurements from within the secondary shear zone were around 363 ± 8 HV. However, it should be noted that the measurements were approximately 20 μm from the outer edge of the chip which was as near to the cutting interface as could be reliably tested through microhardness measurements.

To gain a higher level understanding of the deformation microstructures formed during machining, TEM investigations were undertaken on samples from the expanded shear band regions formed in the primary deformation zone and from the secondary shear zone, shown in Figures 3 and 4, respectively.

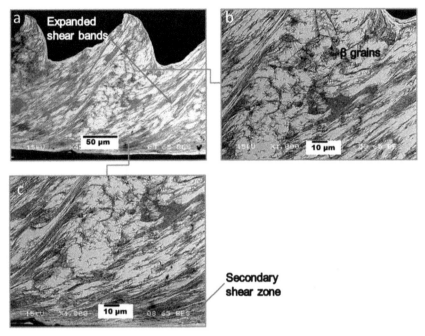

Figure 2. SEM of machining chip microstructure for surface cutting speed around 90 m/min: (**a**) Low magnification image of segmented chip morphology. (**b**) Undeformed chip region with original β grain structure of the workpiece material with in-tact β grains highlighted. (**c**) Higher magnification image of the secondary shear zone which adjoins the tool rake face during cutting.

3.3. Transmission Electron Microscopy (TEM) Analysis of Expanded Shear Band Region

A montage of bright field (BF) TEM images showing the typical microstructure within the expanded shear band region is presented in Figure 3a and a selected area diffraction pattern (SADP) from this region is included inset. They reveal that the microstructure in the bands is highly refined, consisting of fine elongated β grains as the matrix phase interspersed with lath-like α precipitates (lighter contrast). The SAPD, indexed to the β and α phases, exhibits a continuous ring pattern characteristic of very fine and randomly oriented grains with large grain boundary misorientations. There is a gradient of deformation from the highly refined, smaller grains at the centre to more coarse grains at the left and right extremities. Some individual β grains can be identified (arrowed in Figure 3a) due to dark strain contrast arising from high levels of dislocation activity. The arrowed β grains reveal the progression of refinement from the extremities to the interior. The β grains at the extremities have a diamond-like shape with grain sizes in the order of 400–500 nm with α phase laths often sitting along their diagonal boundaries. Closer to the heavily refined central region, the β grains transition to an elongated form with grains around 20–50 nm in width and 300–400 nm in length, while the α laths with their long axis closely aligned to the length-wise axis of the β grains are 5–10 nm in width and 300–400 nm in length. In the most highly refined region, the β grain size is less than 100 nm.

Higher magnification BF and corresponding hollow cone dark field (HCDF) images from this highly refined region are presented in Figure 3b,c, respectively. They further reveal the fine β grains with considerable internal deformation structure as well as the lath-like α phase. A high magnification BF image showing the coarse diamond shaped β grains at the extremities of the deformation region is shown in Figure 3d.

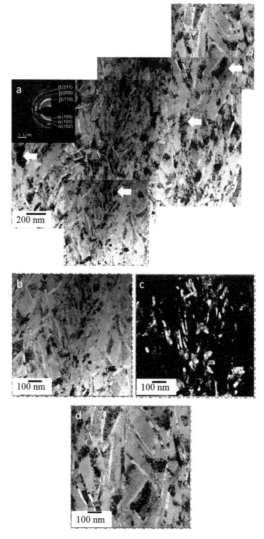

Figure 3. Transmission electron microscopy (TEM) from the expanded shear band region: (**a**) Montage of bright field (BF) images with selected area diffraction pattern (SADP) inset indexed to the β and α phases. (**b**) BF image and (**c**) corresponding hollow cone dark field (HCDF) image from the red dashed region in (**a**) formed from β (110) and α (100), (002), (101) diffraction rings. (**d**) Higher magnification BF image from the blue dashed region in (**a**).

3.4. TEM Analysis of Secondary Shear Zone

A series of TEM images from the secondary shear zone are presented in Figure 4. They reveal the chip microstructure in the region immediately adjacent to the rake face (a), i.e., at the chip extremity, and then moving incrementally into the chip interior, (b) and (c).

Figure 4a shows a BF TEM image with SADP inset and corresponding HCDF image from the region immediately adjacent to the rake face. The SADP exhibits a continuous ring pattern characteristic of very fine and randomly oriented grains with reflections from the β phase only. The BF image reveals a gradient of deformation from left to right from fine equiaxed grains immediately adjacent to the cutting interface at the left with grain sizes around 30–50 nm to more elongated grains 50 to 100 nm in width and several hundreds of nm in length at the right further from the cutting interface. A BF TEM image showing the significantly elongated β grains formed further from the cutting interface (approximately 1–2 μm) and a corresponding HCDF formed from the β (110) diffraction ring are shown in Figure 4b. Figure 4c reveals the microstructure in the secondary shear zone 5–10 μm from the cutting interface becomes less refined and consists of significantly larger, elongated β grains with long lath-like α precipitates (lighter contrast). Again, some β grains exhibit dark contrast arising from high levels of dislocation activity within their interiors. The SAPD from this region exhibits a discontinuous ring pattern indicative of a less refined structure with reflections from both the β and α phases.

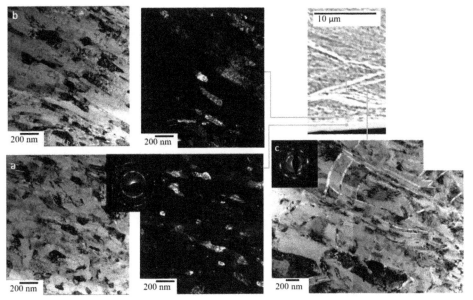

Figure 4. Series of TEM images from the secondary shear zone region immediately adjacent to the rake face (**a**), i.e., the chip extremity, and incrementally further into the chip interior (**b**,**c**). (**a**) BF image with SADP inset showing reflections almost entirely from the β phase and corresponding HCDF image formed from the β (110) diffraction ring. (**b**) BF image and corresponding HCDF image formed from β (110) diffraction ring. (**c**) Montage of BF images with SADP inset with reflections from the β and α phases.

4. Discussion

The TEM characterisation revealed important details of the machining chip microstructures which are linked to the deformation processes taking place in the chips during machining of the Ti-25Nb-3Mo-3Zr-2Sn alloy. Significant differences were identified between the expanded shear bands regions and the secondary shear zone which are discussed in respect to the influence of strain, heat generation and temperatures encountered during cutting.

High levels of deformation in the expanded shear bands formed within the primary deformation zone promote extensive localised refinement of the β phase to almost equiaxed grains with sizes below 100 nm. Additionally, α phase laths frequently occupy the β grain boundaries, aligned to the shear direction. This grain refinement consequently led to an approximate 50% increase in hardness from that of the starting material, primarily due to locally enhanced Hall-Petch strengthening [18]. In regions adjacent to the fine equiaxed zone, there is a transition to firstly larger elongated β grains, which according to the work of Zhan et al. [16] can be inferred to be the <001> and <111> fibre textures, with α laths aligned to the shear direction, and then to larger diamond shaped β grains with α phase often located on their diagonal axes. Across the entire width of the expanded shear band region, this variation in refinement suggests a cyclic process of formation such that the expanded shear band consists of an accumulation of individual localised shear events. The presence of both β and α phases across the entire expanded shear band region indicates that temperatures associated with adiabatic heating during cutting do not exceed the β transus temperature of around 720 °C for the Ti-25Nb-3Mo-3Zr-2Sn alloy [19]. The high density of α phase laths in this region suggest that dynamic precipitation may also take place in conjunction with the deformation. This suggests that temperatures during cutting are likely in the range of 400–600 °C [20].

Shear localisations are often observed in titanium alloys subject to dynamic loading associated with their low heat conductivity and high adiabatic shearing sensitivity [21]. Chip formation occurs by concentrated shear within the deformation band and the microstructural refinement is attributed to the large shear strains imposed. The formation of the shear regions is linked to substantial reductions in the yield stress which occur during cutting due to substantial thermal softening of the Ti-25Nb-3Mo-3Zr-2Sn alloy at elevated temperatures [16]. Under the momentum diffusion-based shear band evolution model, a highly localised primary shear band forms at the centre of the primary deformation zone and large deformation occurs inside the shear band as the localised shear deformation proceeds [22]. Subsequent thermal softening enables relaxation of the stress within the shear band and the stress relaxation further propagates into the surrounding undisturbed material giving rise to momentum diffusion and broadening of the deformation region as the expanded shear band regions evolve.

In comparison to the highly localised shear bands observed in cutting of other titanium alloys [23–26], the expanded shear band regions observed for the Ti-25Nb-3Mo-3Zr-2Sn alloy are relatively diffuse. Under the current cutting conditions, for the Ti-25Nb-3Mo-3Zr-2Sn alloy these zones are around 100 μm in width and of a similar magnitude to that of the regions with more limited deformation. An increase in hardness of around 10% from the starting material in regions between the expanded shear bands indicates that deformation in the primary shear zone is not entirely confined to the expanded shear band regions. Additionally, increased temperatures during cutting may also promote further α phase precipitation in these regions of more limited deformation which would also increase hardness.

For the secondary shear zone, strains at the cutting interface can be very large (>5) [27], typically leading to formation of ultrafine and/or nanocrystalline grain structures. The level of strain and hence refinement decays with increasing distance from the cutting interface across the secondary shear zone, which is in the order of 15–20 μm in width for the Ti-25Nb-3Mo-3Zr-2Sn alloy. At the cutting interface the β grains are heavily refined and approximately equiaxed with very fine nanocrystalline grain sizes around 20–50 nm while further from the interface, approximately 1–2 μm, the β grains become highly elongated in the shear direction with grains in the order of 100 nm in width and 0.5 to 1 μm in length. Again it can be inferred that the elongated grains are aligned to the <001> and <111> fibre textures [16]. An absence of the α phase in the equiaxed and elongated β grain regions of the secondary shear zone indicates that recrystallization takes place during cutting and temperatures are in excess of the alloy's 720 °C β transus temperature while subsequent rates of cooling are sufficiently high to preserve the single phase β microstructure. At 5–10 μm from the cutting interface a mixture of larger elongated β grains and long α laths are observed and the hardness is enhanced by more than 30% over that of the starting material. This can be attributed to the effects of significant β phase refinement resulting from deformation during cutting which enhances Hall-Petch strengthening as well as dynamic precipitation taking place due to the elevated temperatures during cutting. The α laths tend to have thinner, longer morphologies than those observed within the expanded shear band region which is potentially due to the influence of comparatively higher temperatures and strains in this region, which favour growth of precipitates over nucleation through dynamic precipitation effects.

As the temperature and intensity of strain decrease across the secondary shear zone as a function of the distance from the cutting interface, the microstructures observed reflect their differing thermomechanical histories. The variations observed in the microstructure across the secondary deformation zone for the Ti-25Nb-3Mo-3Zr-2Sn alloy are largely consistent with those reported for adiabatic shear bands formed in titanium alloys [21,28,29] involving a progression to finer, more equiaxed grains in the region of most intense deformation, in this case at the cutting interface [14].

Previously, Rashid et al. [12] used infrared thermography to measure the maximum temperatures at the back surface of the chips in the cutting region during machining of the Ti-25Nb-3Mo-3Zr-2Sn alloy. They reported the average temperatures to be 540–600 °C for surface cutting speeds of 80–100 m/min for the solution treated and aged alloy. However, as the cutting edge was covered by the chip, the temperatures reported were those from the cutting zone at the external surface of the chip, i.e., on the opposite side to the cutting interface. Therefore, the actual temperature at the interface between the cutting face of the tool and the chip may be substantially higher than those reported. Additionally, infrared thermography temperature measurements are acknowledged to be subject to substantial error [30,31] and temperatures in the cutting zone are far from uniform, being typically characterised by regions of high temperature gradient [32,33]. The microstructural characterisation of the chips gives some insight into the degree of these temperature gradients for machining of the solution treated and aged Ti-25Nb-3Mo-3Zr-2Sn alloy. For the chip, the significant sources of heat are in the primary deformation zone due to plastic work associated with shear and in the secondary deformation zone due to work done in deformation of the chip and in association with sliding friction at the tool-chip interface.

Based on the assumption that all mechanical work is converted to heat, the cutting forces can be applied to estimate heat generation during cutting. In this case, the heat generated in the primary deformation zone, Q_s, can be calculated from [32]:

$$Q_s = W_c = F_V \cdot V \tag{1}$$

where F_V is the tangential cutting force and V is the cutting velocity.

The amount of heat generated due to work done in the secondary deformation zone along the tool rake face is calculated from the frictional energy given by:

$$Q_r = \frac{F_{fr} \cdot V}{\lambda} \tag{2}$$

where F_{fr} is the total shear force acting on the rake face and λ is the chip thickness ratio. The shear force can be calculated from:

$$F_{fr} = F_V \sin(\alpha) + F_S \cos(\alpha) \tag{3}$$

where F_S is the feed force and α is the rake angle. An estimate of the heat generated in the primary and secondary deformation zone made from the above equations using the cutting forces reported by Rashid et.al. [12] with a tangential cutting force, $F_V = 450$ N, a feed force, $F_S = 240$ N, a cutting velocity, $V = 90$ m/min, a chip thickness ratio, $\lambda = \frac{0.19 \text{ mm}}{0.15 \text{ mm}}$ and a rake angle $\alpha = 9°$ .yields heat generation of approximately 40 kW in the primary deformation zone and 22 kW in the secondary shear zone. From this analysis, the heat generated in the secondary shear zone for cutting of the Ti-25Nb-3Mo-3Zr-2Sn alloy is proportionately quite high at around 55% of that in the primary deformation zone. For context, Tay et al. observe that typically the total heat generation due to plastic deformation and frictional sliding in the secondary deformation zone for continuous chips from a non-abrasive material at medium cutting speeds is around 20% to 30% of the heat generated in the primary cutting zone [34].

While cutting heat is removed by the chip, the tool, the workpiece and some of the heat generated at the shear plane (primary shear zone) is transferred to the tool-chip interface. Hence, the temperature in the chip in the region adjacent to the tool rake face rises due to the combination of heat from the primary and secondary shear zones. Predictions of the temperature fields in the chip on the basis of heat generation are complex and have been the subject of various analytical and numerical investigations involving modelling of heat conduction, kinematics, geometries and energy aspects of the machining process. Others have attempted to measure temperature both at the cutting interface zone and across the chip, tool and workpiece through the methods including embedded thermocouples, radiation pyrometers and metallographic techniques [32,33]. In general, the highest temperatures are reportedly near the tool-chip interface in the secondary deformation zone. This is consistent with the microstructural analysis of the Ti-25Nb-3Mo-3Zr-2Sn alloy chips which indicated that temperatures in this region were in excess of 720 °C during cutting.

5. Conclusions

Ultrafine grain microstructures formed by severe plastic deformation during machining of the solution treated and aged Ti-25Nb-3Mo-3Zr-2Sn biomedical β titanium alloy have been investigated by TEM analyses of specimens obtained by FIB from transverse section of chips. The investigations have revealed that:

1. High levels of deformation in the primary shear zone promote extensive refinement of the β phase within expanded shear band regions approximately 100 μm in width to almost equiaxed grains with sizes below 100 nm in regions of intense deformation, while α phase laths frequently occupy the grain boundaries aligned to the shear direction. There is a transition to firstly elongated β grains and then to larger diamond-shaped β grains in adjoining regions of less intense deformation. The presence of a high density of α phase laths across the entire expanded shear band region suggests that temperatures in this region are likely in the range of 400–600 °C during cutting.

2. For the secondary shear zone, large strains at the cutting interface result in recrystallised, approximately equiaxed grains with nanocrystalline grain sizes around 20–50 nm, while further (1–2 μm) from the interface the β grains become highly elongated in the shear direction with grains in the order of 100 nm in width and 0.5 to 1 μm in length. At the cutting interface, an absence of the α phase indicates that the temperatures exceed the alloy's 720 °C β transus temperature. At 5–10 μm, from the cutting interface a mixture of large elongated β grains and long α phase laths are observed. The microstructural variation across the secondary shear zone reflects the decay of strain and temperature away from the cutting interface.

3. The microstructural characterisation of the chips infers information on the temperature fields present across the chips during cutting. The highest cutting temperatures occur within the secondary shear zone at the cutting interface, associated with proportionately high levels of heat generation due to deformation and friction.

Author Contributions: Conceptualization, D.K., M.D., S.S. and M.B.; Methodology, D.K.; Investigation, D.K., R.R.R. and H.A.; Writing-Original Draft Preparation, D.K.; Writing-Review & Editing, R.R.R., M.B. and S.S.; Project Administration, M.D.; Funding Acquisition, D.K. M.B. and M.D.

Funding: This research was funded by the Australian Research Council (ARC) Research Hub for Advanced Manufacturing of Medical Devices (IH150100024).

Acknowledgments: The authors acknowledge the facilities, and the scientific and technical assistance of the Australian Microscopy & Microanalysis Research Facility at the Centre for Microscopy and Microanalysis, The University of Queensland and at the Australian Centre for Microscopy & Microanalysis at the University of Sydney.

Conflicts of Interest: The authors declare no conflict of interest.

References

1. Long, M.; Rack, H.J. Titanium alloys in total joint replacement—A materials science perspective. *Biomaterials* **1998**, *19*, 1621–1639. [CrossRef]

2. Niinomi, M. Mechanical biocompatabilities of titanium alloys for biomedical applications. *J. Mech. Behav. Biomed. Mater.* **2008**, *1*, 30–42. [CrossRef] [PubMed]

3. Ping, D.H.; Mitarai, Y.; Yin, F.X. Microstructure and shape memory behavior of a Ti-30Nb-3Pd alloy. *Scr. Mater.* **2005**, *52*, 1287–1291. [CrossRef]

4. Yu, Z.; Wang, G.; Ma, X.Q.; Dargusch, M.S.; Han, J.Y.; Yu, S. Development of biomedical near beta titanium alloys. In *Materials Science Forum: 4th International Light Metals Technology Conference*; Trans Tech Publications: Zurich, Switzerland, 2009.

5. Kent, D.; Wang, G.; Yu, Z.; Dargusch, M.S. Pseudoelastic behaviour of a β Ti-25Nb-3Zr-3Mo-2Sn alloy. *Mater. Sci. Eng. A* **2010**, *527*, 2246–2252. [CrossRef]

6. Kent, D.; Wang, G.; Yu, Z.; Ma, X.; Dargusch, M.S. Strength enhancement of a biomedical titanium alloy through a modified accumulative roll bonding technique. *J. Mech. Behav. Biomed. Mater.* **2011**, *4*, 405–416. [CrossRef] [PubMed]

7. Rahman, M.; Wong, Y.S.; Zareena, A.R. Machinability of titanium alloys. *JSME Int. J. Ser. C* **2003**, *46*, 107–115. [CrossRef]

8. Yang, X.; Liu, C.R. Machining titanium and its alloys. *Mach. Sci. Technol.* **1999**, *3*, 107–139. [CrossRef]

9. Ezugwu, E.O.; Wang, Z.M. Titanium alloys and their machinability—A review. *J. Mater. Process. Technol.* **1997**, *68*, 262–274. [CrossRef]

10. Machado, A.R.; Wallbank, J. Machining of titanium and its alloys—A review. *Proc. Inst. Mech. Eng. Part B* **1990**, *204*, 53–60. [CrossRef]

11. Uzer, B.; Toker, S.M.; Cingoz, A.; Bagci-Onder, T.; Gerstein, G.; Maier, H.J.; Canadinc, D. An exploration of plastic deformation dependence of cell viability and adhesion in metallic implant materials. *J. Mech. Behav. Biomed. Mater.* **2016**, *60*, 177–186. [CrossRef] [PubMed]

12. Rashid, R.A.R.; Sun, S.; Wang, G.; Dargusch, M.S. Machinability of a near beta titanium alloy. *Proc. Inst. Mech. Eng. Part B* **2011**, *225*, 2151–2162. [CrossRef]

13. Rashid, R.A.R.; Sun, S.; Wang, G.; Dargusch, M.S. Experimental investigation of various chip parameters during machining of the Ti25Nb3Mo3Zr2Sn beta titanium alloy. *Adv. Mat. Res.* **2013**, *622*, 366–369. [CrossRef]

14. Schneider, J.; Dong, L.; Howe, J.Y.; Meyer, H.M. Microstructural characterization of Ti-6Al-4V metal chips by focused ion beam and transmission electron microscopy. *Metall. Mater. Trans. A* **2011**, *42*, 3527–3533. [CrossRef]

15. Zhan, H.; Zeng, W.; Wang, G.; Kent, D.; Dargusch, M. On the deformation mechanisms and strain rate sensitivity of a metastable β Ti-Nb alloy. *Scr. Mater.* **2015**, *107*, 34–37. [CrossRef]

16. Zhan, H.; Wang, G.; Kent, D.; Dargusch, M. The dynamic response of a metastable β Ti-Nb alloy to high strain rates at room and elevated temperatures. *Acta Mater.* **2016**, *105*, 104–113. [CrossRef]

17. Ye, G.G.; Xue, S.F.; Ma, W.; Dai, L.H. Onset and evolution of discontinuously segmented chip flow in ultra-high-speed cutting Ti-6Al-4V. *Int. J. Adv. Manuf. Technol.* **2017**, *88*, 1161–1174. [CrossRef]

18. Hughes, G.D.; Smith, S.D.; Pande, C.S.; Johnson, H.R.; Armstrong, R.W. Hall-petch strengthening for the microhardness of twelve nanometer grain diameter electrodeposited nickel. *Scr. Metall.* **1986**, *20*, 93–97. [CrossRef]

19. Zhentao, Y.; Lian, Z. Influence of martensitic transformation on mechanical compatibility of biomedical β type titanium alloy tlm. *Mater. Sci. Eng. A* **2006**, *438*, 391–394. [CrossRef]

20. Kent, D.; Pas, S.; Zhu, S.; Wang, G.; Dargusch, M.S. Thermal analysis of precipitation reactions in a Ti–25nb–3mo–3zr–2sn alloy. *Appl. Phys. A* **2012**, *107*, 835–841. [CrossRef]

21. Wang, B.; Wang, X.; Li, Z.; Ma, R.; Zhao, S.; Xie, F.; Zhang, X. Shear localization and microstructure in coarse grained beta titanium alloy. *Mater. Sci. Eng. A* **2016**, *652*, 287–295. [CrossRef]

22. Ye, G.G.; Xue, S.F.; Jiang, M.Q.; Tong, X.H.; Dai, L.H. Modeling periodic adiabatic shear band evolution during high speed machining Ti-6Al-4V alloy. *Int. J. Plast.* **2013**, *40*, 39–55. [CrossRef]

23. Arrazola, P.J.; Garay, A.; Iriarte, L.M.; Armendia, M.; Marya, S.; Le Maître, F. Machinability of titanium alloys (Ti6Al4V and Ti555.3). *J. Mater. Process. Technol.* **2009**, *209*, 2223–2230. [CrossRef]

24. Joshi, S.; Pawar, P.; Tewari, A.; Joshi, S.S. Effect of β phase fraction in titanium alloys on chip segmentation in their orthogonal machining. *CIRP J. Manuf. Sci. Technol.* **2014**, *7*, 191–201. [CrossRef]

25. Sun, S.; Brandt, M.; Dargusch, M.S. Characteristics of cutting forces and chip formation in machining of titanium alloys. *Int. J. Mach. Tool. Manuf.* **2009**, *49*, 561–568. [CrossRef]

26. Dargusch, M.S.; Sun, S.; Kim, J.W.; Li, T.; Trimby, P.; Cairney, J. Effect of tool wear evolution on chip formation during dry machining of ti-6al-4v alloy. *Int. J. Adv. Manuf. Tech.* **2018**, *126*, 13–17. [CrossRef]

27. Oxley, P.L.B. *Mechanics of Machining*; Ellis Horwood: New York, NY, USA, 1989.

28. Zhan, H.; Zeng, W.; Wang, G.; Kent, D.; Dargusch, M. Microstructural characteristics of adiabatic shear localization in a metastable beta titanium alloy deformed at high strain rate and elevated temperatures. *Mater. Charact.* **2015**, *102*, 103–113. [CrossRef]

29. Yang, Y.; Jiang, F.; Zhou, B.M.; Li, X.M.; Zheng, H.G.; Zhang, Q.M. Microstructural characterization and evolution mechanism of adiabatic shear band in a near beta-ti alloy. *Mater. Sci. Eng. A* **2011**, *528*, 2787–2794. [CrossRef]

30. Davies, M.A.; Ueda, T.; M'Saoubi, R.; Mullany, B.; Cooke, A.L. On the measurement of temperature in material removal processes. *CIRP Ann. Manuf. Technol.* **2007**, *56*, 581–604. [CrossRef]

31. Lane, B.; Whitenton, E.; Madhavan, V.; Donmez, A. Uncertainty of temperature measurements by infrared thermography for metal cutting applications. *Metrologia* **2013**, *50*, 637–653. [CrossRef]

32. Abukhshim, N.A.; Mativenga, P.T.; Sheikh, M.A. Heat generation and temperature prediction in metal cutting: A review and implications for high speed machining. *Int. J. Mach. Tool. Manuf.* **2006**, *46*, 782–800. [CrossRef]

33. Sutter, G.; Ranc, N. Temperature fields in a chip during high-speed orthogonal cutting—an experimental investigation. *Int. J. Mach. Tool. Manuf.* **2007**, *47*, 1507–1517. [CrossRef]

34. Tay, A.O.; Stevenson, M.G.; De Vahl Davis, G. Using the finite element method to determine temperature distributions in orthogonal machining. *Proc. Inst. Mech. Eng.* **1974**, *188*, 627–638. [CrossRef]

 *metals*

Article

Incremental Forming of Titanium Ti6Al4V Alloy for Cranioplasty Plates—Decision-Making Process and Technological Approaches

Sever Gabriel Racz, Radu Eugen Breaz *, Melania Tera, Claudia Gîrjob, Cristina Biriş, Anca Lucia Chicea and Octavian Bologa

Department of Industrial Machines and Equipment, Engineering Faculty, "Lucian Blaga" University of Sibiu, Victoriei 10, 550024 Sibiu, Romania; gabriel.racz@ulbsibiu.ro (S.G.R.); melania.tera@ulbsibiu.ro (M.T.); claudia.girjob@ulbsibiu.ro (C.G.); cristina.biris@ulbsibiu.ro (C.B.); anca.chicea@ulbsibiu.ro (A.L.C.); octavian.bologa@ulbsibiu.ro (O.B.);
* Correspondence: radu.breaz@ulbsibiu.ro; Tel.: +40-745-374-776

Received: 4 July 2018; Accepted: 6 August 2018; Published: 9 August 2018

Abstract: Ti6Al4V titanium alloy is considered a biocompatible material, suitable to be used for manufacturing medical devices, particularly cranioplasty plates. Several methods for processing titanium alloys are reported in the literature, each one presenting both advantages and drawbacks. A decision-making method based upon AHP (analytic hierarchy process) was used in this paper for choosing the most recommended manufacturing process among some alternatives. The result of AHP indicated that single-point incremental forming (SPIF) at room temperature could be considered the best approach when manufacturing medical devices. However, Ti6Al4V titanium alloy is known as a low-plasticity material when subjected to plastic deformation at room temperature, so special measures had to be taken. The experimental results of processing parts from Ti6Al4V titanium alloy by means of SPIF and technological aspects are considered.

Keywords: single-point incremental forming; AHP; cranioplasty plates; decision-making; titanium alloys; medical devices

---

1. Introduction

Titanium alloys are considered eligible materials for biomedical applications (implants and prosthetic devices) due to their biocompatibility. The work presented in [1] provides a comprehensive analysis regarding the main types of titanium alloys used in biomedical applications, as well as their advantages and main drawbacks. A review regarding the titanium alloys seen as the best solution for orthopedic implants is presented in [2], where also the main requirements for a material to be considered a biomaterial are introduced. One of the requirements for this is biocompatibility, which according to [3] is measured by how the human body reacts to the device made of this material when it is implanted. In this work, hip and knee implants are defined as the main orthopedic implants. Both studies presented in [1] and [2] mention Ti6Al4V alloy as one of the titanium alloys; it was initially developed for the aeronautical industry, but can be successfully used for biomedical applications.

Cranioplasty is another main field where titanium alloys may be used due to their biocompatibility. A detailed review about the techniques and materials used in cranioplasty is presented in [4]. Titanium is considered one of the suitable materials, being biocompatible but hard to shape. Another review [5] also indicated titanium alloys as one the materials of choice for cranioplasty plates.

The work presented in [6] reported the successful use of 300 plates of titanium for cranioplasty. The requested shape of the cranioplasty plate was determined either by a traditional technique or by means of computer tomographic scans. Finally, the plates were shaped by pressing them against a counter-die, which could be considered a plastic deformation process. A comprehensive study presented in [7] confirmed titanium alloys as one of the recommended materials for cranioplasty plates, but also highlighted the fact that complications occurred in 29% out of 127 cases. However, the recommendations were to strengthen the prophylaxis measures against infections, rather than replacing titanium as the material for cranioplasty plates. The research was not focused upon the method of manufacturing titanium cranioplasty plates.

Another study reporting the use of titanium plates for cranioplasty was presented in [8]. The work was towards using CAD (computer aided design)/CAM (computer aided manufacturing) methods for manufacturing the plates. A rapid prototyping method based upon fine casting was used and it was reported as having several advantages compared with a traditional milling process.

Titanium alloys were considered the best choice for cranioplasty of large skull defects, according to the results presented in [9]. The study was based upon long-term observations of 26 patients and emphasized the fact that none of the titanium plates implanted had to be removed. Even if the study mentioned CAD/CAM techniques for manufacturing cranioplasty plates, these techniques were not described in detail, and were mostly oriented on the generation of the requested shape of the plate using computer tomography, rather than presenting how the plates were manufactured.

The approaches regarding the methods of manufacturing the titanium plates for cranioplasty are very diversified. The study presented in [10] emphasized the advantages of a manual approach (the shape of the plate was obtained by pressing the titanium sheet against a template model using a manual press), while in [11] the shape of the plate was obtained by means of multiforming, a method which requires very complex technological equipment with a high degree of automatization.

Thus, it can be concluded that titanium alloys are suited for manufacturing cranioplasty plates and there is no consecrated technological approach, either manual or automated, for that. Finding a suitable method for manufacturing the plates was one of the objectives of this work and it involves, in the first stage, a review of the main methods of manufacturing parts from titanium alloys.

The work presented in [12] emphasizes the effects of machining Ti6Al4V alloy by means of cutting, using tools made of straight-grade cemented carbide. Microstructure alterations were reported and, moreover, the reported surface roughness falls into the rough machining category. Consequently, cutting processes may not be the recommended solution when machining biomedical devices from titanium alloys.

Usually, the titanium alloys are also low-plasticity materials, so processing them by means of plastic deformation is also difficult. Single-point incremental forming (SPIF) is one of the manufacturing processes used for processing titanium alloys to overcome the drawback introduced by the low plasticity of these materials, which prevents their processing by other plastic deformation processes. A schematic diagram of the SPIF process is presented in Figure 1, where the sheet metal workpiece (2) is fixed by means of the retaining plate (3) and active plate (backing plate) (1). The punch (4) is moving in the vertical direction, along the Z axis, while the assembly formed by (1)–(3) executes a movement in the XY plane. By combining these movements, various trajectories can be achieved and, subsequently, different shapes of the sheet metal final part.

Literature reviews regarding the SPIF process are presented in [13], which covers the results obtained before 2005, and most recently in [14], which synthesizes the research results reported between 2005 and 2015. The influence of various SPIF process parameters upon the results is synthesized in [15]. It is here noticeable the fact that the maximum achievable angle for a truncated cone part made of Ti6Al4V alloy processed by SPIF using a 10 mm diameter punch reported in [14] was 32°, the lowest one compared with all materials processed by means of SPIF. For comparison, for similar geometry and tool but for DC04 steel and AA 5754 (AlMg3) aluminum alloy, the reported maximum wall angles were 64° and 71°, respectively. These results stress the fact that special measures must be taken when unfolding the SPIF process using titanium alloys as workpieces.

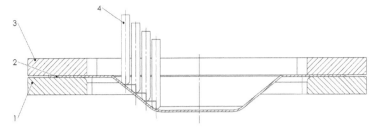

Figure 1. Schematic diagram of asymmetric single-point incremental forming (SPIF).

A comprehensive study about machining commercially pure titanium (CP Ti) by means of incremental forming is presented in [16]. The experiments have proven that by a proper selection of tool (diameter, material) and lubricant (type and lubrication method), wall angles up to 65° can be achieved for CP Ti. Certain values for the ratio between tool diameter (d) and processing pitch (p) were recommended (d/p ≤ 40) for better results with regards to surface quality.

High feed rates and rotation speeds were tested for parts processed by SPIF and promising results were presented in [17], but the experiments took place on aluminum alloys. Moreover, high feed rates were used for reducing the processing time, in contrast with the results shown in [18], which indicated that formability is inversely proportional with the feed rate. Another experimental study using aluminum alloys for the test parts presented in [19] indicated that the use of high rotation speeds for the forming tools can improve the formability by lowering the forming force. On the other hand, the surface roughness is improved by using the punch rotation, while the rotating speed does not influence it. In [20], Titanium grade 2 and Ti6Al4V were machined on a CNC (computer numerically controlled) lathe using high feed rates. The results have shown that high-speed SPIF does not adversely affect the microstructure of the materials. However, the study was not focused upon the formability, and the geometrical shape of the part was a cone frustum with a wall angle of 25° (for the Ti6Al4V), which was considered by the authors as safe from this point of view.

A new technique of SPIF which involves the heating of the machined workpiece was proposed in [21]. Sheet metal workpieces made of AZ31 magnesium and TiAl2Mn1.5, both with low formability at room temperature, were processed. The parts were heated using direct current (DC) with values between 300 and 600 A, and good results were reported for machining symmetrical parts (cone frustum) from a formability point of view. However, because the method is subject to a patent it was not clearly described how the temperature was controlled and, moreover, how the heat did affect the microstructure of the materials, which is very important when considering the biocompatibility. Another work, presented in [22], used also heat supplied by means of DC to machine several materials including Ti6Al4V by means of SPIF. The studies have reported an increase in the formability (a maximum wall angle of 35° for the cone frustum part made of Ti6Al4V), but also microstructure alterations in the form of different grain distributions were observed. The roughness of the machined part also increased with the increase of the wall angle. Another approach, presented in [23], combined the local heating with high tool speed to machine a car body element made of Ti6Al4V alloy by means of SPIF. The studies have indicated that at 400 °C, the formability of the parts increases, while the normal anisotropy was not influenced by the temperature.

Local heating of the workpiece by a laser beam system, coaxial with the punch, integrated in the main spindle system of the CNC machine-tool used for the SPIF process, was presented in [24]. The path for the laser beam was calculated by taking into consideration the part geometry. Improvements with regards to formability were reported on parts made of Ti6Al4V alloy, where the maximum machining depths were higher than those obtained at room temperature. No references were made with regards to temperature measurement and control or the influence of the heat upon the microstructure of the machined materials. Another method for applying heat on parts machined by SPIF was presented in [25], where friction obtained by tool rotation was used. At speeds between 2000 and 7000 rpm, the heat generated by friction improved the formability of the parts, but as pointed out by the authors it is not only due to material softening, but also to recrystallization. However, the material used for this research was AA5052-H32 aluminum alloy, so no information about using frictional heat for machining titanium alloys by means of SPIF was available.

A master–slave tool layout for double-side incremental forming (DSIF) combined with electrically assisted heating was presented in [26]. Good results were reported in processing lightweight materials (AZ31B magnesium alloy) with regards to surface quality and maximum wall angle for truncated cone parts and a new type of hybrid toolpath lead to better geometric accuracy. However, the machining layout for DSIF, which must be custom-built, and the necessity to control and synchronize the toolpaths of the master and slave tools could lead to very high machining costs in contrast to the real value of the machined parts.

With regards to the influence of high temperatures applied during manufacture processes upon the biocompatibility of titanium alloys, there are still many opinions. The biocompatibility of these materials is related to the spontaneous formation of a passive oxide layer at room temperature, which is reported to reduce oxygen diffusion and further oxidation at lower temperatures [27–29]. However, at higher temperatures the situation is changing. The works presented in [30–32] consider that oxidation at high temperatures limits the applications of these kind of alloys. In [30] is stated the fact that diffusion of oxygen at temperatures above 400 °C leads to the development of a hard and brittle oxygen diffusion zone which leads to a loss of tensile ductility and of fatigue resistance, reducing the life expectancy of titanium alloys. Above 600 °C, a thick and defective oxide layer is formed, facilitating the penetration of oxygen into the material [30]. Aluminum was found to diffuse outwards through the oxide layer in the later oxidation stage [33]. A study reported in [29] stated that the presence of aluminum in outer layers of Ti6Al4V alloy may hinder osteointegration (bone bonding to the implant) when used as an implant material. Also, aluminum is known to cause neurological disorders [34].

However, a controlled oxidation process called "thermal oxidation" at high temperature is also seen as a promising technique to improve protection against friction and wear [30,32,35–37].

There are also arguments which support the fact that forming titanium alloys at high temperature does not affect their compatibility. Promising results regarding the manufacturing of cranioplasty plates by SPIF with material heating were presented in [38]. The workpiece was heated at 650 °C during the SPIF process and impact tests were performed to measure the maximum force and energy absorbed by the plate. Furthermore, a cytotoxicity test was also performed to assess if the manufacturing process affected the biocompatibility of the plate. The test showed no differences between the processed surfaces and the control ones with regards to biocompatibility. The work presented in [39] had shown that the influence of oxygen enrichment during the process of manufacturing cranial prostheses by means of superplastic forming did not affect the biocompatibility of the Ti6Al4V alloy. A cytotoxicity test was performed, and the viability of the cells was not affected.

Consequently, it is not yet fully demonstrated if heating the material during the process does or does not affect its biocompatibility. However, processing at high temperatures significantly improves the plasticity of the titanium alloys. On the other hand, the complexity of the equipment, the costs associated with that complexity (and with higher energy consumption), and the difficulty of controlling the process favor processing at room temperature, if plasticity requirements can be fulfilled.

From an environmental and sustainability point of view, recent works have pointed out that SPIF is a process with higher amounts of energy consumption compared with other forming processes, such as stamping [40]. The studies presented in [41] have indicated tool speed, type of material, and vertical incremental step in this order as the main influence factors upon the amount of energy consumption during the SPIF process. A similar study which also took into consideration the technological equipment (CNC machine, six-axes industrial robot, and dedicated Amino machine) was presented in [42] and emphasized the fact that forming time is the most influential factor upon the electric energy consumption. Based upon this assumption, the study presented in [43] compared the energy efficiency of performing SPIF machining on a CNC milling machine and on a high-speed CNC lathe. The results have demonstrated that high-speed processing significantly reduces the processing time and, consequently, the energy consumption. Environmental aspects must be considered every time machining is involved, but medical devices such as implants and prosthetic devices are usually machined as prototypes; thus, the environmental impact manufacturing them should be considered quite low. However, reducing the overall machining time is one of the most recommended approaches from this point of view.

The authors of this work have performed some previous studies with regards to using complex trajectories and computer-assisted techniques in SPIF processes [44,45] and some preliminary work in the field of using titanium alloys in cranial implants [46].

As presented above, there are many techniques in use for manufacturing cranioplasty plates from Ti6Al4V alloy, each one with advantages and drawbacks. However, as presented in the next section, single-point incremental forming at room temperature could be considered the best choice, if some criteria are considered. Of course, as reported in the literature, the low plasticity of the Ti6Al4V alloy makes it difficult to process it in this condition. Thus, the approach presented in this paper was oriented towards finding some technological approaches which could allow for better processing of Ti6Al4V alloy at room temperature. As results, some incremental findings, linked mainly with the types of toolpaths and the values of processing steps, with regards to the proposed objective were synthesized.

2. Materials and Methods

2.1. Decision-Making Process

Cranioplasty plates may be manufactured using either a manual or a digital approach (Figure 2). The input data come either from a physical template (a bone fragment taken form the patient) or from a computer tomography (CT) scan. If the manual manufacturing approach is chosen, by means of the physical template, a negative cast is made using laboratory putty. The following steps involve several manual operations, which may differ from one method to another. A comprehensive description of a manual method for manufacturing cranioplasty plates is presented in [10].

The digital approach relies heavily on CAD (computer-aided design)/CAM (computer-aided manufacturing) techniques. The data from the CT scan is processed and converted from a point cloud model to a 3D STL (stereolithography) file format by means of any CAD software package. The 3D model is imported into any CAM software where processing technology and machining code are generated and sent to the technological equipment (CNC machine tool, industrial robot, or even a specialized machine). The format of the machining code and the type of the technological equipment depend on the technological process used for the actual manufacturing of the cranioplasty plate. No matter the chosen manufacturing method, in the final stage, the cranioplasty plate is subjected to some specific operations to prepare it for implantation (i.e., sterilization).

According to the literature review presented above, the most recommended manufacturing processes for cranioplasty plates are cutting (CUT), single-point incremental forming (SPIF), single-point incremental forming with heating (SPIFH), and double-side incremental forming (DSIF). A special mention should be made with regards to additive manufacturing methods, which were recently reported as effective methods for manufacturing cranioplasty plates. An approach presented in [47] presented a two-stage method for manufacturing a cranioplasty plate. In the first stage, by means of 3D printing, a mold was manufactured which was further used for casting a polymethyl-methacrylate (PMMA) cranioplasty plate in the second stage. More recently, an industrial company presented a case study [48] in which a cranioplasty plate was manufactured using a metal 3D printing machine, using Ti MG1 (ISO 10993) as the material.

However, the current research is oriented upon using Ti6Al4V as the material for manufacturing a cranioplasty plate; thus, only CUT, SPIF, SPIFH, and DSIF will be considered for the analysis.

A decision-making method for selecting between these processes based upon AHP (analytic hierarchy process) was developed during this research.

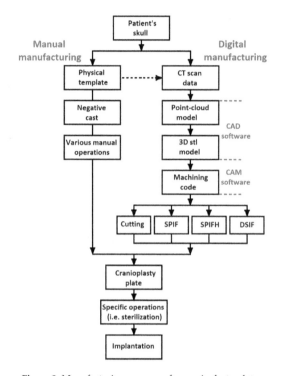

Figure 2. Manufacturing processes for cranioplasty plates.

Comparing the four considered manufacturing processes is a multiattribute decision-making problem due to the factors involved. One of the methods which can be used for this purpose is the Analytic Hierarchy Process (AHP), a method introduced by Saaty [49,50]. The method is based upon pairwise comparison. Elements *i* and *j* are compared, and the result is expressed by value a_{ij}. A given hierarchization criteria is used for the comparison:

$$a_{ij} = 1 \quad for \ i = j, \quad where \ i,j = 1, 2, \ldots n$$
$$a_{ij} = \tfrac{1}{a_{ji}} \quad for \ i \neq j \tag{1}$$

The judgement scale, used for AHP, was proposed by Saaty: 1, equally important; 3, weakly more important; 5, strongly more important; 7, demonstrably more important; 9, absolutely more important. The values in between (2, 4, 6, and 8) represent compromise judgements.

To use the AHP process for comparing the four manufacturing processes, a set of seven criteria were proposed and compared pairwise against each other. The preference matrix from Table 1 is used to store the results. The six proposed criteria are presented below:

- C1—Formability: seen here as the ability of the manufacturing process to modify the shape of the workpiece by redistributing the material (plastic deformation). It is noticeable here that three of the analyzed processes are plastic deformation processes, while one of them (CUT) is based upon shaping the part by removing material. However, it was considered that this criterion could be also applied to the CUT process;

- C2—Microstructure: seen here as a measure of how the microstructure of the material is affected by the manufacturing process and, consequently, how the biocompatibility of the processed part could be affected;

- C3—Degree of control: seen here as a measure of how the parameters of the process and the shape and dimensional parameters of the parts (cranioplasty plates) can be controlled;

- C4—Roughness: the meaning of this criterion is quite straightforward, as it expresses the surface quality achievable for the processed parts;

- C5—Energy consumption: it is related with the amount of energy required by each manufacturing process;

- C6—Accuracy: seen here as the maximum achievable accuracy for the parts processed by each of the analyzed manufacturing processes;

- C7—Production time: seen here as the total amount of time to produce a cranioplasty plate.

Table 1. Preference matrix A.

Criterion	C1	C2	C3	C4	C5	C6	C7
C1	1	1/3	5	3	7	3	5
C2	3	1	9	3	9	5	3
C3	1/5	1/9	1	1/5	3	1/5	1/5
C4	1/3	1/3	5	1	7	5	1/3
C5	1/7	1/9	1/3	1/7	1	1/7	1/7
C6	1/3	1/5	1/5	1/5	7	1	1/3
C7	1/5	1/3	5	3	7	3	1

As an example, the way in which the first line of Table 1 was filled is presented below:

- Microstructure (C2) and formability (C1) are very important characteristics of a cranioplasty plate; however, for a device in contact with the human tissue, the state of the microstructure should be considered weakly more important that the ability of shaping the plate;

- The degree of control (C3) is a measure of the quality and repeatability of the process. A higher degree of control will allow the process to be automated, but, finally, for a prosthetic device (which can also be manufactured manually), the ability to shape the plate exactly as required (C1) is strongly more important;

- Roughness (C4) of the part is also important for a prosthetic device, but while the microstructure cannot be repaired if affected by the manufacturing process, roughness could be improved (even by manual operations); thus, the formability of the plate (C1) should be considered weakly more important;

- Energy consumption (C5) should be reduced as possible for any manufacturing process; however, when it comes to cranioplasty plates (which usually are manufactured as prototypes), the ability of shaping the part should (C1) be considered demonstrably more important than saving energy (C5);

- Manufacturing accuracy of the cranioplasty plate (C6) is important, but from the point of view of its functional role (prosthetic device, which is not moving or being in contact with other moving parts), the formability (C1) should be considered weakly more important;

- Production time (C7) is a measure of the efficiency of a production process, but taking into consideration of the fact that, as stated for the (C5) criterion, the cranioplasty plates are manufactured as prototypes, the (C1) criterion should be considered strongly more important.

The next step of the AHP process involves the normalization of the preference matrix by transforming it into matrix B, where

$$B = [b_{ij}]$$
$$b_{ij} = \frac{a_{ij}}{\sum_{i=1}^{n} a_{ij}} \qquad (2)$$

It is now required to calculate the eigenvector $w = [w_i]$, which expresses the preference between the elements, by using the following relationship:

$$w_{ij} = \frac{\sum_{i=1}^{n} b_{ij}}{n} \qquad (3)$$

The normalized B is presented in Table 2. The eigenvector w was placed on the last column of matrix B, calculated using Equation (3).

Table 2. Normalized matrix B.

Criterion	C1	C2	C3	C4	C5	C6	C7	w
C1	0.1996	0.1458	0.3024	0.3842	0.2188	0.2901	0.2239	0.2256
C2	0.5989	0.4375	0.4234	0.3842	0.2188	0.2901	0.3134	0.3905
C3	0.0399	0.0625	0.0605	0.0427	0.0938	0.0193	0.1343	0.0558
C4	0.0665	0.1458	0.1815	0.1281	0.2188	0.2901	0.1343	0.1777
C5	0.0285	0.0625	0.0202	0.0183	0.0313	0.0138	0.0149	0.0233
C6	0.0665	0.1458	0.0121	0.0427	0.2188	0.0967	0.1343	0.0829
C7	0.0384	0.0640	0.0160	0.0423	0.0811	0.0227	0.0448	0.0442

The comparisons must be checked from the point of view of consistency, according to [48–51]. The check is made by calculating the maximal eigenvalue according to

$$\lambda_{max} = \frac{1}{n} \sum_{i=1}^{n} \frac{(Aw)_i}{w_i} = 7.4469 \qquad (4)$$

where λ_{max} is the matrix's largest eigenvalue [34].

Using the random consistency index table (Table 3) from [50], the consistency ratio CR may be determined (for a 6-dimensional matrix, the r coefficient is 1.32).

Table 3. Values for CI indices.

Size of Matrix (n)	1	2	3	4	5	6	7	8	9	10
Random average CI (r)	0	0	0.58	0.90	1.12	1.24	1.32	1.41	1.45	1.51

According to Equation (5), the value of CR is smaller than 10%, showing that the comparisons made during the building of matrices A and B are consistent [36,37].

$$CR = \frac{\lambda_{max} - n}{r(n-1)} = 5.64\% \qquad (5)$$

The evaluation of the four manufacturing strategies with respect to the seven criteria will be unfolded below. The evaluation for each criterion is presented in Tables 4–10, together with the eigenvectors (introduced in the last column of each table). For exemplification, the way in which the second line of Table 4 was filled is presented below:

- Cranioplasty plates are manufactured starting from a sheet metal workpiece; thus, a plastic deformation process (SPIF) should be considered as an intermediate between equally important and weakly more important than a cutting process (CUT) from the point of view of formability (C1). Even the workpiece is different for cutting, and cutting also allows the user to machine complex shapes; thus an intermediate value has been considered;

- Ti6Al4V alloy is known as a low-formability material, and heating it leads to an increase in the formability. However, applying heat could lead to some problems described above. Thus, SPIFH should be considered weakly more important than SPIF, from the (C1) point of view;
- Using a master–slave tools layout with punch and counter-punch will significantly improve the formability of the part, but will also lead to the use of very complex layouts and equipment; this is why DSIF should be considered weakly more important than SPIF, from the (C1) point of view.

Table 4. Comparison of the processing strategies with regards to C1 (formability).

C1	CUT	SPIF	SPIFH	DSIF	w
CUT	1	1/2	1/2	1/2	0.1386
ASPIF	2	1	1/3	1/3	0.1622
ASPIFH	2	3	1	1/2	0.2902
DSPIF	2	2	2	1	0.4090

Table 5. Comparison of the processing strategies with regards to C2 (microstructure).

C2	CUT	SPIF	SPIFH	DSIF	w
CUT	1	1/9	1/5	1/7	0.0399
ASPIF	9	1	7	7	0.6440
ASPIFH	5	1/7	1	1/3	0.1145
DSPIF	7	1/7	3	1	0.2016

Table 6. Comparison of the processing strategies with regards to C3 (degree of control).

C3	CUT	SPIF	SPIFH	DSIF	w
CUT	1	3	5	5	0.5143
ASPIF	1/3	1	5	5	0.3045
ASPIFH	1/5	1/5	1	3	0.1158
DSPIF	1/5	1/5	1/3	1	0.0654

Table 7. Comparison of the processing strategies with regards to C4 (roughness).

C4	CUT	SPIF	SPIFH	DSIF	w
CUT	1	1/7	1/5	1/7	0.0328
ASPIF	7	1	5	1/3	0.3520
ASPIFH	5	1/5	1	7	0.3199
DSPIF	7	3	1/7	1	0.2953

Table 8. Comparison of the processing strategies with regards to C5 (energy consumption).

C5	CUT	SPIF	SPIFH	DSIF	w
CUT	1	3	5	5	0.4941
ASPIF	1/3	1	9	7	0.3713
ASPIFH	1/9	1/5	1	1/2	0.0528
DSPIF	1/7	1/5	2	1	0.0818

Table 9. Comparison of the processing strategies with regards to C6 (accuracy).

C6	CUT	SPIF	SPIFH	DSIF	w
CUT	1	7	5	3	0.5761
ASPIF	1/7	1	1/2	1/3	0.0715
ASPIFH	1/5	2	1	1/3	0.1125
DSPIF	1/3	3	3	1	0.2399

Table 10. Comparison of the processing strategies with regards to C7 (production time).

C7	CUT	SPIF	SPIFH	DSIF	w
CUT	1	5	5	7	0.5430
ASPIF	1/5	1	3	5	0.2445
ASPIFH	1/5	1/3	1	3	0.0765
DSPIF	1/7	1/5	1/3	1	0.1360

The matrix C will be built using the results from Tables 4–10. The columns of matrix C represent the eigenvectors resulting by comparing the four processes pairwise. The order of the columns within matrix C takes into consideration the order of the criteria determined in Table 2: C2, C1, C4, C6, C3, C7, and C5. Performing the multiplication of matrix C and the vector w, the preference vector **x** for the four manufacturing strategies may be obtained, according to the following relation:

$$x = Cw = \begin{bmatrix} 0.0399 & 0.1386 & 0.0328 & 0.5761 & 0.5143 & 0.5967 & 0.4941 \\ 0.6440 & 0.1622 & 0.3520 & 0.0715 & 0.3045 & 0.2292 & 0.3713 \\ 0.1145 & 0.2902 & 0.3199 & 0.1125 & 0.1158 & 0.0188 & 0.0528 \\ 0.2016 & 0.4090 & 0.2953 & 0.2399 & 0.0654 & 0.0553 & 0.0818 \end{bmatrix} \times \begin{bmatrix} 0.2256 \\ 0.3905 \\ 0.0558 \\ 0.1777 \\ 0.0233 \\ 0.0829 \\ 0.0442 \end{bmatrix} = \begin{bmatrix} 0.2506 \\ 0.2835 \\ 0.1919 \\ 0.2740 \end{bmatrix} \quad (6)$$

As can be noticed from Equation (6), the resulting column matrix has the highest value on the second line, 0.2835, a position which corresponds to the second analyzed manufacturing strategy, (ASPIF). According to this result, the AHP process has returned ASPIF as the most recommended approach if the seven proposed criteria are considered. Consequently, the experimental program was oriented to this process as the preferred solution for manufacturing cranioplasty plates. To assess the robustness and the reliability of the AHP process results, a sensitivity analysis was introduced according to the method proposed in [52,53]. According to this, the weights were changed while maintaining the ranking order previously determined. According to the proposed method, a coefficient $\alpha \geq 0$ is introduced and the matrix A is transformed into $\left[a_{ij}^{\alpha}\right]$. If $\alpha > 1$, more dispersed weights are obtained and if $\alpha < 1$, the weights become more concentrated, without any change in the ranking order.

Table 11 shows the weights obtained for $\alpha = 0.5, 0.7, 0.9, 1.0, 1.1, 1.3, 1.5$ (values proposed in [52]). Table 12 presents the simulation results of calculating the preference vector x for the weights from Table 11. A graphical synthesis of the sensitivity analysis is presented in Figure 3. It can be noticed that the changes in the weights do not affect the hierarchy of the preference vectors x; consequently, SPIF is the most recommended process for the entire range of the analysis.

Table 11. Sensitivity analysis for the weights.

	Coefficient α						
Criterion	0.5	0.7	0.9	1.0	1.1	1.3	1.5
C1	0.1128	0.15792	0.20304	0.2256	0.24816	0.29328	0.3384
C2	0.19525	0.27335	0.35145	0.3905	0.42955	0.50765	0.58575
C3	0.0279	0.03906	0.05022	0.0558	0.06138	0.07254	0.0837
C4	0.08885	0.12439	0.15993	0.1777	0.19547	0.23101	0.26655
C5	0.01165	0.01631	0.02097	0.0233	0.02563	0.03029	0.03495
C6	0.04145	0.05803	0.07461	0.0829	0.09119	0.10777	0.12435
C7	0.0221	0.03094	0.03978	0.0442	0.04862	0.05746	0.0663

Table 12. Results of the sensitivity analysis simulations for the preference vector x.

	Coefficient α/Preference Vector x						
Strategy	**0.5**	**0.7**	**0.9**	**1.0**	**1.1**	**1.3**	**1.5**
CUT	0.1253	0.1754	0.2255	0.2506	0.2757	0.3258	0.3759
SPIF	0.1417	0.1984	0.2551	0.2835	0.3118	0.3685	0.4252
SPIFH	0.0959	0.1343	0.1727	0.1919	0.2111	0.2495	0.2878
DSPIF	0.1370	0.1918	0.2466	0.2740	0.3014	0.3562	0.4110

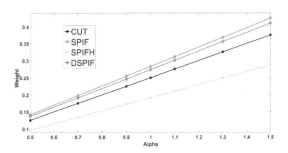

Figure 3. Graphical synthesis of the sensitivity analysis.

2.2. Experimental Layout

As stated in the literature review, one of the most used types of technological equipment for ASPIF processing are CNC milling machines. For this research, a Haas MiniMill CNC machining center was used. The milling machine and the experimental layout mounted on the machine are presented in Figure 4.

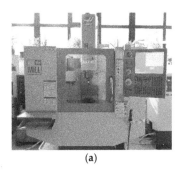 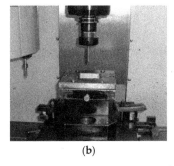

(a)   (b)

Figure 4. Technological equipment: (**a**) Haas MiniMill computer numerically controlled (CNC) machining center; (**b**) Forming equipment (active die and retaining plate).

Figure 5 presents a 3D model of the forming equipment, where 1, punch; 2, active die; 3, retaining plate; 4, active die support; 5, baseplate; 6, fixing screws; 7, centering screw; and 8, sheet metal workpiece.

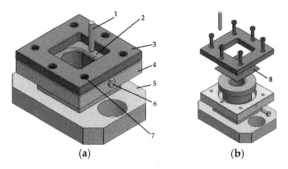

| (a) | (b) |

Figure 5. Forming equipment: (**a**) 3D model of the forming assembly—without the sheet metal workpiece; (**b**) Exploded view—with the sheet metal workpiece.

The active plate used for processing the parts is presented in Figure 6. Both the active die and the punch were made from 20Cr115 SR EN ISO 4957:2002 alloyed steel, heat treated.

Figure 6. Active die used for manufacturing the test parts.

2.3. Material

The chemical composition of Ti6Al4V titanium alloy in mass percentage is shown in Table 13.

Titanium is an allotropic substance consisting of a cubic structure (α-Ti) and a compact hexagonal structure up to a temperature of 882 ° C (β-Ti). As can be seen in Table 13, the main alloying elements are aluminum and vanadium, but besides these there are also other minor alloying elements such as iron, oxygen, nitrogen, hydrogen, silicon, and so on.

Table 13. Alloying elements of the Ti6Al4V material.

Alloy Element	Chemical Symbol	Mass Percentage (%)
Aluminum	Al	5.5–6.75
Vanadium	V	3.5–4.5
Carbon	C	0.10
Iron	Fe	0.3
Oxygen	O	0.02
Nitrogen	N	0.05
Hydrogen	H	0.015
Silicon	Si	0.15
Remainder	-	0.4

The mechanical characteristics of the titanium alloy Ti6Al4V are given in Table 14.

Table 14. Mechanical characteristics of the titanium alloy Ti6Al4V.

Characteristic	Measurement Unit	Value
Yield Strength	[MPa]	965–1103
Tensile Strength	[MPa]	896–1034
Density	[g/cm^3]	4.5
Modulus of Elasticity (Young modulus)	[GPa]	116

To determine the other mechanical characteristics of the material needed for finite element method (FEM) analysis, the tensile test was used. The tests were carried out for Ti6Al4V titanium alloy with a thickness of 0.5 mm, using the following laboratory equipment:

- tensile testing machine Instron 5587;
- optical strain measurement system GOM Aramis.

One of the methods of testing the deformation is the uniaxial traction test. On this machine, the specimen is fixed at both ends and deformed at a constant speed until cracking occurs.

Test specimens used for tensile testing are specimens with a calibrated length of 75 mm, a width of 12.5 mm, and a rectangular cross section (Figure 7) in accordance with the standard for the traction testing of metallic materials, SR EN 10002-1: 2002.

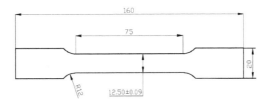

Figure 7. Specimens used for tensile testing.

To study the material anisotropy, sets of specimens were cut (by waterjet cutting) at 0°, 45°, and 90° angles to the sheet rolling direction; these are shown in Figure 8.

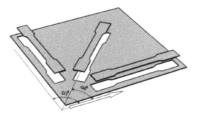

Figure 8. The different angles of the specimens.

The parameters related to the intrinsic properties of the material measured by traction test are hardening coefficient, coefficient of resistance, and coefficients of plastic anisotropy. The values will be used to define the elastoplastic behavior of the material in the FEM simulation. The tensile tests were performed on 3 sets of samples at room temperature, according to Table 15.

Table 15. Tensile tests.

No.	No. of Specimens/Set	Direction of Lamination (°)	Temperature (°C)
1.	3	90°	25 °C
2.	3	0°	25 °C
3.	3	45°	25 °C

Using the BlueHill version 2.0 software (produced by Instron company, Norwood, MA, USA) to control the Instron 5587 Traction Testing Machine (produced also by Instron company), the following were set as input data: type of test, initial dimensions of the specimen, and deformation speed. Both BlueHill software and Instron machine are in the laboratories of Lucian Blaga University of Sibiu.

Following the data processing, the conventional strain curves (σ) versus elongation (ε_{max}) were obtained for the titanium alloy Ti6Al4V at room temperature, which are shown in Figure 9.

The mechanical characteristics of the material that were determined by the traction test are

- modulus of elasticity E [MPa],
- flow limit $R_{p0.2}$ [MPa],
- tensile strength R_m [MPa],
- hardening coefficient n [-],
- resistance coefficient K [Pa],
- elongation ε_{max} [%].

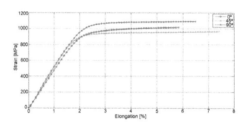

Figure 9. Conventional strain curves (σ) versus elongation (ε_{max}).

In Table 16, a synthesis of the data obtained for the tensile testing for the three types of samples is presented.

Table 16. Synthesis of the data obtained from tensile tests.

Characteristic	Measurement Unit	Value		
Specimen Cutting Angle	[°]	0	45	90
The modulus of elasticity E	[MPa]	49,645.24	49,779.71	52,587.8
Flow Limit $R_{p0.2}$	[MPa]	881.9	863.11	922.51
Tensile Strength R_m	[MPa]	960.76	992.76	1001.35
Coefficient of hardening n	[-]	0.16	0.1	0.13
Resistance coefficient K	[Pa]	1618.9	1190.88	1392.81
Elongation ε_{max}	[%]	5.5	7.8	6.4

2.4. Shape of Test Parts

To test the proposed technological approach, a truncated cone shape of the part was chosen. The geometry of the test part was defined by the cone angle ($\alpha°$), the height of the part (h), and the diameter of the upper base (d). The shape and dimensions of the parts are presented in Figure 10.

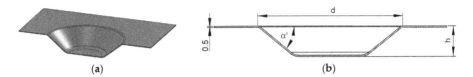

Figure 10. Test parts: (**a**) 3D model; (**b**) Characteristic dimensions.

2.5. Processing Trajectories

The processing trajectories were selected by taking into consideration the third objective stated above. According to the literature review, two main solutions have been imposed lately:

- contour-curves-based trajectory (a contour curve is obtained by intersecting the 3D shape by an XY plane—for the truncated cone, the contour curve is a circle);
- spatial spiral trajectory.

The trajectories used during the experimental test are synthesized in Table 17 and Figure 11.

Table 17. Trajectories.

Trajectory Type	Geometrical Primitive	Code	Observations
Circular trajectories	Contour curve (circle)	CT	The lead-in/lead-out points are lying on the same line (cone generatrix)
Circular trajectories with special entry points	Contour curve (circle)	CTSEP	The lead-in/lead-out points are distributed on the part surface
Spiral trajectories	Spatial spiral	ST	Only one lead-in and one lead-out point

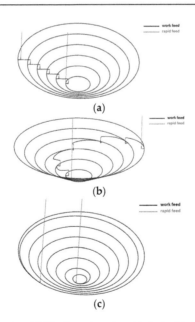

Figure 11. Processing trajectories: (**a**) Circular trajectories with the lead-in/lead-out points lying on the same line; (**b**) Circular trajectories with the lead-in/lead-out points distributed on the surface; (**c**) Spiral trajectories with only one lead-in and one lead-out point.

A computer-aided manufacturing software package, SprutCAM v. 11 (produced by Sprut Technology Ltd., Naberezhnye Chelny, Russia) dedicated for milling operations, was used for generating the trajectories. By using a commercially available CAM solution, the goal regarding ease of generation has been reached.

The circular trajectories (CT) (Figure 11a) have the drawback that all the lead-in points are situated on the same line, a cone generatrix. Lead-in points are the points where the tool (punch) enters in contact with the part. For the circular trajectories, the punch approaches the part with rapid feed, changes it in work feed in the near vicinity of the part, and enters in contact with the workpiece, all these movements being unfolded on the Z axis. After that, the relative movement between the punch and workpiece is unfolded in the XY plane, until a full circle is completed. After completing the circle, the punch performs a new lead-in movement combined on the XY plane and Z axis, and engages the part on a new circle, situated at distance p from the first one, where p is the vertical step of the ASPIF process. In Figure 11a, all the lead-in points are situated on the same line, a situation which can lead to stress accumulation and, finally, to cracks. It is here noticeable that for a CAM solution for milling, aligning the lead-in points is a default procedure.

To avoid this drawback, the second approach was used. In the circular trajectories with special entry points (CTSEP) situation (Figure 11b), the lead-in points were distributed on the lateral surface of the truncated cone. To achieve this distribution, approach and retraction paths in the XY plane had to be defined (Figure 12) where 1, workpiece; 2, approach path in the XY plane; 3, start of the contour curve (circle); 4, retraction path in the XY plane; and 5, finish of the contour curve (circle).

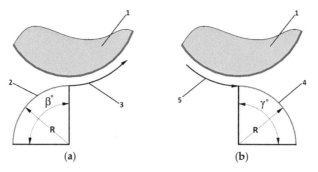

Figure 12. Approach and retraction: (**a**) Approach path in the XY plane defined by radius R and angle β; (**b**) Retraction path in the XY plane defined by radius R and angle γ.

The following relations for the values R, β, and γ were used:

$$\beta = \gamma = 45°$$ (6)

By combining the approach and retraction paths in the XY plane with the approach and retraction paths on the Z axis, the movement cycle of the punch may be divided as follows (Figure 13):

- the punch approaches on the Z axis with rapid feed (a);
- the punch continues the approach on the Z axis with work feed, until it reaches the contour curve level (b);
- the punch follows the approach path, in an XY plane at the Z level of the contour curve, until it is positioned on the contour curve (c);
- the punch follows the contour curve (d);
- the punch follows the retraction path, in the XY plane situated at the Z level of the contour curve (e);
- the punch approaches on the Z axis with rapid feed, travelling to the next contour curve (f);

- the punch continues the approach on the Z axis with work feed, until it reaches the next contour curve level—a new XY plane (g);
- after phase (g), the movements are repeating in a cycle, from d to g, until the last contour curve is processed.

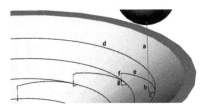

Figure 13. Movement phases for circular trajectories with special entry points (CTSEP).

By performing the movement phases described in Figure 12, the lead-in and lead-out points are distributed on the lateral surface of the cone, thus avoiding the accumulation of stresses and consequently avoiding the occurrence of cracks.

The third approach uses a spatial spiral trajectory (Figure 11c). In this case the trajectory is a continuous one, with only one entry point (lead-in) and one exit point (lead-out).

2.6. Processed Parts

The experimental tests were conducted according to the following parameters:

- The working feedrate was fixed to 150 mm/min;
- The punch was fixed in the main spindle of the machine and driven with a rotational speed of 150 rev/min around its own axis. According to the literature review, this rotation reduces the friction and has a favorable influence upon the formability of the material. However, the rotational speed was limited to avoid the heating of the material, which could affect its formability. The temperature limit in this case is 400 °C. At 150 rev/min, the temperature (measured during the process with an FLIR TermoVision A320 thermal imaging camera (manufactured by FLIR Systems, Inc., Wilsonville, OR, USA) was found to be lower than 100 °C;
- The starting angle of the truncated cone was set to 30°, the next one was 35°, and afterwards the angle was incremented by 1°;
- Three vertical steps were considered: 0.2, 0.4, and 0.6 mm. Smaller steps, i.e., 0.1 mm were considered too small to be considered from a technological point of view, while steps greater than 0.6 mm lead to crack occurrence events at an angle of 30°;
- Two punch diameters were considered: 8 and 10 mm;
- Mineral oil was used as lubricant;
- At each angle, the first approach involved the use of the simplest trajectory (CT). If for a given angle this trajectory failed (crack occurrence), the CTSE was used instead. If the latter failed also, the ST trajectory was considered;
- The experimental tests are synthesized in Table 18. The lines in Table 18 only present the parts which were processed without cracks (successful tests). Each successful test was confirmed by performing it three times.

Table 18. Synthesis of the experimental tests.

Crt. No.	Base Diameter d [mm]	Vertical Step p [mm]	Height h [mm]	Cone Angle α [°]	Punch Diameter d_p [mm]	Trajectory Type
1.		0.4		30	8	CT
2.		0.4		30	10	CT
3.		0.6		30	8	CT
4.		0.6		30	10	CT
5.		0.2		35	8	CT
6.		0.2		35	10	CT
7.		0.4		35	8	CTSEP
8.		0.4		35	10	CTSEP
9.		0.6		35	8	CTSEP
10.		0.6		35	10	CTSEP
11.		0.2		36	8	CT
12.	55	0.2	12	36	10	CT
13.		0.4		36	8	CTSEP
14.		0.4		36	10	CTSEP
15.		0.6		36	8	ST
16.		0.6		36	10	CTSEP
17.		0.2		37	8	CTSEP
18.		0.2		37	10	CTSEP
19.		0.4		37	8	ST
20.		0.4		37	10	ST
21.		0.6		37	10	ST
22.		0.2		38	8	ST
23.		0.2		38	10	ST
24.		0.4		38	10	ST

Some of the successful tests are presented in Figures 14–16.

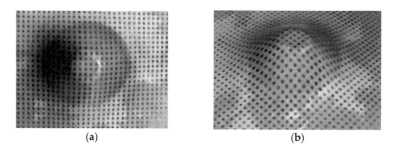

Figure 14. Part with p = 0.2 mm, α = 30°, d_p = 8 mm, and CT. (**a**) view from above; (**b**) side view.

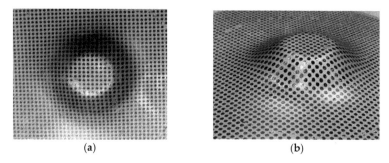

Figure 15. Part with p = 0.6 mm, α = 35°, d_p = 10 mm, and CTSEP. (**a**) view form above; (**b**) side view.

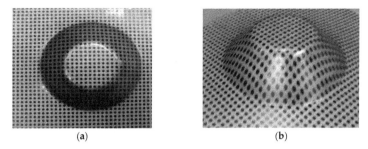

Figure 16. Part with $p = 0.4$ mm, $\alpha = 38°$, $d_p = 10$ mm, and ST. (**a**) view from above; (**b**) side view.

Some examples of processed parts which cracked during the ASPIF process are presented in Figures 17–19.

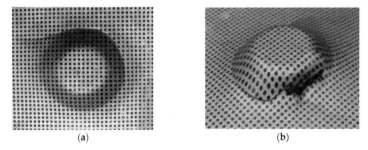

Figure 17. Part with $p = 0.6$ mm, $\alpha = 35°$, $d_p = 8$ mm, and CT. (**a**) view from above; (**b**) side view.

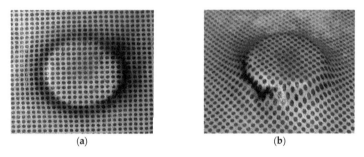

Figure 18. Part with $p = 0.4$ mm, $\alpha = 38°$, $d_p = 8$ mm, and ST. (**a**) view from above; (**b**) side view.

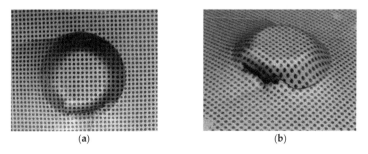

Figure 19. Part with $p = 0.4$ mm, $\alpha = 40°$, $d_p = 10$ mm, and ST. (**a**) view from above; (**b**) side view.

To test the accuracy of the processed parts, some measurements were performed using a Mahr profilometer (from Mahr Gmbh, Göttingen, Germany). Figure 20 presents a graphical display of the measurement results for Part 24, while Table 19 presents a synthesis of the results for the measured parts. Even if more dimensional characteristics were measured, the results were focused on the angle $\alpha°$ and on the surface roughness (expressed by Ra and Rz). It is here noticeable the fact that taking into consideration the functional role intended for the parts (cranioplasty plates), their requested accuracy lies in a very different range compared with parts from the manufacturing industry, for example.

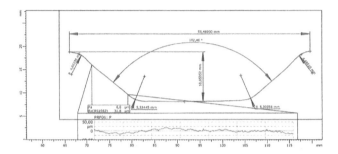

Figure 20. Measurement results for Part 24.

Table 19. Synthesis of results for the measured parts.

Part No.	Characteristics	Measured Values
1.	$\alpha° = 30/p = 0.4/d_p = 8$ mm/CT	$\alpha° = 31.485$ Ra = 15.4 μm/Rz = 23.1 μm
3.	$\alpha° = 30/p = 0.6$ mm/$d_p = 8$ mm/CT	$\alpha° = 31.125$ Ra = 19 μm/Rz = 52.4 μm
6.	$\alpha° = 35/p = 0.2$ mm/$d_p = 10$ mm/CT	$\alpha° = 36.125$ Ra = 6 μm/Rz = 35.3 μm
8.	$\alpha° = 35/p = 0.4$ mm/$d_p = 10$ mm/CTSEP	$\alpha° = 35.86$ Ra = 16.5 μm/Rz = 58.2 μm
11.	$\alpha° = 36/p = 0.2$ mm/$d_p = 8$ mm/CT	$\alpha° = 35.865$ Ra = 10.3 μm/Rz = 37.5 μm
18.	$\alpha° = 37/p = 0.2$ mm/$d_p = 10$ mm/CTSEP	$\alpha° = 38.8$ Ra = 8.3 μm/Rz = 21 μm
23.	$\alpha° = 38/p = 0.2$ mm/$d_p = 10$ mm/ST	$\alpha° = 38.56$ Ra = 3.5 μm/Rz = 18.3 μm
24.	$\alpha° = 38/p = 0.4$ mm/$d_p = 10$ mm/ST	$\alpha° = 38.77$ Ra = 6.8 μm/Rz = 31.6 μm

2.7. FEM Analysis

The Abaqus/Explicit software package, v.14 (produced by Dassault Systèmes®, Vélizy-Villacoublay, France) as used for the FEM analysis. A parameterized model based upon the play between the punch and the active plate, the retention pressure, the diameter of the blank, and the radius of the active plate was developed. The geometric model included the sheet metal workpiece (considered as a deformable body), the active plate, the retention plate, and the punch (all being considered as rigid bodies). For the finite element mesh, four-node shell elements were used. The modeling was done on medium fiber, with five integration points per thickness being considered. The FEM model is presented in Figure 21.

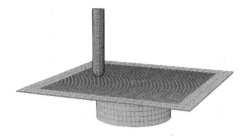

Figure 21. FEM model.

The parameters targeted by the FEM simulations were:

- major strains (ε_1);
- minor strains (ε_2);
- thickness reduction (s_{max});
- evolution of the forming force on Z axis.

Figures 22–24 present the variations in ε_1, ε_2, and s_{max} for a truncated cone with diameter of the upper base d = 55 mm, cone angle α = 30°, vertical step p = 0.4 mm, punch diameter d_p = 10 mm, and spatial spiral trajectories.

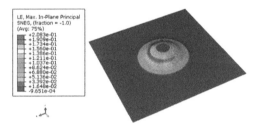

Figure 22. Distribution of major strains (ε_1)—diameter of the upper base d = 55 mm, cone angle α = 30°, vertical step p = 0.4 mm, punch diameter d_p = 10 mm, and spatial spiral trajectories (ST).

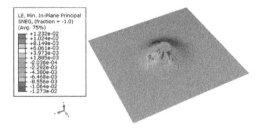

Figure 23. Distribution of minor strains (ε_2)—diameter of the upper base d = 55 mm, cone angle α = 30°, vertical step p = 0.4 mm, punch diameter d_p = 10 mm, and spatial spiral trajectories (ST).

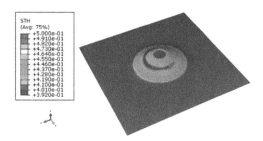

Figure 24. Thickness reduction (s_{max})—diameter of the upper base d = 55 mm, cone angle α = 30°, vertical step p = 0.4 mm, punch diameter d_p = 10 mm, and spatial spiral trajectories (ST).

Figures 25 and 26 present the simulated processing force on the Z axis, for the same part, diameter of the upper base d = 55 mm, cone angle α = 30°, vertical step p = 0.4 mm, and punch diameter d_p = 10 mm, but for different types of trajectories—circles (Figure 22) and spatial spiral (Figure 23). It is here noticeable the fact that the simulation speed was increased by a magnification factor of 100; thus, the time scale covers the whole process.

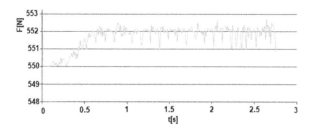

Figure 25. Simulated processing forces on the Z axis—diameter of the upper base d = 55 mm, cone angle α = 30°, vertical step p = 0.4 mm, punch diameter d_p = 10mm, and circular trajectories (CT).

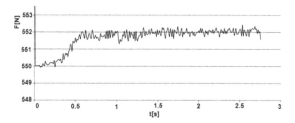

Figure 26. Simulated processing forces on the Z axis—diameter of the upper base d = 55 mm, cone angle α = 30°, vertical step p = 0.4 mm, punch diameter d_p = 10mm, and spatial spiral trajectories (ST).

A preliminary analysis reveals that the maximum values of the forces are quite similar, oscillating around 552 N. However, for the spiral trajectory, the amplitude of the oscillations is higher, a fact that could favor the occurrence of cracks.

2.8. Experimental Measurements

A GOM Argus optical system (produced by GOM company, Braunschweig, Germany) was used for measuring the parts. Figure 27 presents the experimental results for major strain (ε_1) distribution for the part with diameter of the upper base d = 55 mm, cone angle α = 30°, vertical step p = 0.4 mm, punch diameter $d_p = 8$ mm, and circular trajectories (CT). The maximum value of the major strain is 37.6%. Figure 28 presents the results for major strain (ε_1) distribution for the part with $d = 55$ mm, $\alpha = 35°$, $p = 0.4$ mm, $d_p = 8$ mm, and circular trajectories with separate entry points (CTSE). The maximum value of the major strain is 34.1%. Figure 29 presents the results for major strain (ε_1) distribution for the part with $d = 55$ mm, $\alpha = 38°$, $p = 0.4$ mm, $d_p = 10$ mm, and spatial spiral trajectories (ST). The maximum value of the major strain is 23.7%.

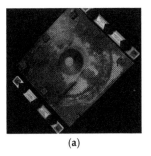
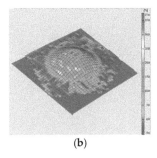

(a)   (b)

Figure 27. Measured distribution of major strains (ε_1)—diameter of the upper base $d = 55$ mm, cone angle $\alpha = 30°$, vertical step $p = 0.4$ mm, punch diameter $d_p = 8$ mm, and circular trajectories (CT), (**a**) measured part; (**b**) measurement results.

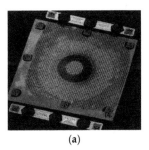
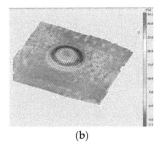

(a)   (b)

Figure 28. Measured distribution of major strains (ε_1)—diameter of the upper base $d = 55$ mm, cone angle $\alpha = 35°$, vertical step $p = 0.4$ mm, punch diameter $d_p = 8$ mm, and circular trajectories with separate entry points (CTSEP), (**a**) measured part; (**b**) measured results.

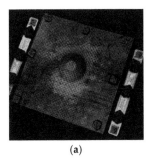
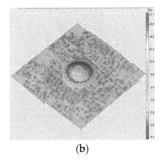

(a)   (b)

Figure 29. Measured distribution of major strains (ε_1)—diameter of the upper base $d = 55$ mm, cone angle $\alpha = 38°$, vertical step $p = 0.4$ mm, punch diameter $d_p = 10$ mm, and spatial spiral trajectories (ST), (**a**) measured part; (**b**) measurement results.

A complete set of measured values of major strains (ε_1), minor strains (ε_2), and thickness reduction (s_{max}) for the part with diameter of the upper base d = 55 mm, cone angle α = 30°, vertical step p = 0.4 mm, punch diameter d_p = 10 mm, and spatial spiral trajectories (ST) is presented in Figure 30.

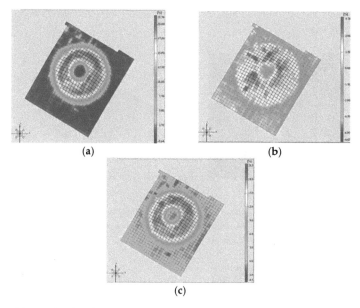

(a)   (b)

(c)

Figure 30. Measured values for major strains (ε_1) (**a**), minor strains (ε_2) (**b**), and thickness reduction (s_{max}) (**c**) for the parts with diameter of the upper base d = 55 mm, cone angle α = 30°, vertical step p = 0.4 mm, punch diameter d_p = 10 mm, and spatial spiral trajectories (ST).

A comparison between the simulated and the experimentally measured values for major strains (ε_1), minor strains (ε_2), and thickness reduction (s_{max}) for the part with diameter of the upper base d = 55 mm, cone angle α = 30°, vertical step p = 0.4 mm, punch diameter d_p = 10 mm, and spatial spiral trajectories (ST) is presented in Table 20.

Table 20. Comparison between simulated and measured values.

Part d = 55 mm, α = 30°, p = 0.4 mm, d_p = 10 mm, ST	Characteristic Input					
	Major Strains ε_1		Minor Strains ε_2		Thickness Reduction s_{max}	
	%	log	%	log	%	log
Experimental	21.52	0.1951	4.78	0.0467	21.5	0.242
Simulated	-	0.2083	-	0.0123	-	0.216

2.9. Manufacturing a Cranioplasty Plate

The next step of the work was to process by means of SPIF a part with complex shape, specific for cranioplasty plates, to demonstrate that the proposed technological conditions allow the user to manufacture irregular shapes with rapid variations of the wall shapes and angles at room temperature. A manually made physical model was considered, taking into consideration the following requirements:

- The shape of the model had to be highly irregular, to mimic as close as possibly the human skull;
- The shape of the model had to present rapid variations of the wall shapes and angles;

- Even if the experimental layout had size limitations, the overall area of the model was chosen about 40 cm² (exactly 36.5 cm²). According to the literature [54,55], this size could be considered as a quite common value for a cranial defect surface area.

After scanning the physical model, the 3D model of the part, presented in Figure 31, was stored in an stl file which resulted after processing a CT scan point cloud file.

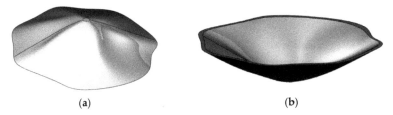

(a)   (b)

Figure 31. 3D model of the cranioplasty plate: (**a**) upper side; (**b**) lower side.

The shape of the plate is highly irregular and continuously variable, as can be seen from Figure 32. However, the wall angles were checked to be lower than 38° for any area of the part.

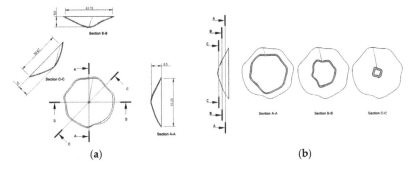

(a)   (b)

Figure 32. Geometry of the plate: (**a**) transversal sections; (**b**) horizontal sections.

Spatial spiral trajectories were used with a vertical step of $p = 0.2$ mm and the punch with $d_p = 10$ mm was chosen as the processing tool. Figure 33 presents the shape of the processing trajectories (toolpath), but for clarity, the vertical step was enlarged ten times (2 mm).

Figure 33. Spiral toolpath (for clarity of presentation, the vertical step was enlarged ten times).

The processed part is presented in Figure 34.

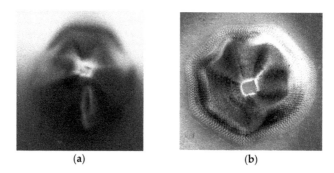

(a)   (b)

Figure 34. Processed part: (**a**) upper side; (**b**) lower side.

Using the Argus GOM optical measurement system, major strains (ε_1), minor strains (ε_2), and thickness reduction (s_{max}) for the cranioplasty plates were measured. A synthesis of the values is presented in Table 21.

Table 21. Measured characteristic values for the cranioplasty plate.

Cranioplasty Plate	Characteristic Input		
	Major Strains ε_1 [%]	Minor strains ε_2 [%]	Thickness Reduction s_{max} [%]
Characteristic	14.5	4.3	17.44

A graphical presentation of the measured values is presented in Figure 35.

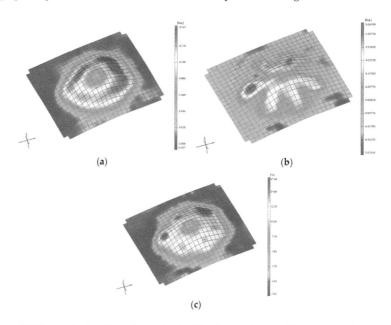

(a)   (b)

(c)

Figure 35. Measured values for major strains (ε_1) (**a**), minor strains (ε_2) (**b**), thickness reduction (s_{max}) (**c**) for the cranioplasty plate.

From Table 21 and Figure 35, it can be noticed that the values of the characteristic inputs are in acceptable ranges for parts manufactured by means of ASPIF. In fact, these values are even smaller than the ones obtained for the test parts (Table 20).

3. Results

The AHP method proposed here has indicated that SPIF is the most recommended manufacturing method if certain criteria are considered. Of course, the method could be affected by subjectivity, and the results could be changed if the analysis is done by other specialists. However, the performed sensitivity analysis guarantees, in some respects, the robustness of the results.

It is here noticeable that AHP indicates the best choice according to the criteria taken into consideration. Thus, if other sets of criteria are chosen, the results may differ significantly. The results presented in this work do not state that SPIF could be considered the best choice in any respect, but it could be considered the best choice if the criteria are those chosen in this approach.

An FEM model was developed which was able to provide simulation results close to values found experimentally for major and minor strains and for thickness reduction.

The experimental program provided some information regarding the technological conditions in which some incremental improvements (mainly an increase in the achievable wall angle) in the results of processing Ti6Al4V alloy at room temperature could be achieved. Also, it was presented that using continuous paths, a part with irregular shapes and with rapid variations of the wall shapes and angles has been processed.

Of course, there are also several shortcomings to this research:

- The processed test parts have limited dimensions (due to the available experimental layout), so it should be demonstrated by further study that the findings are also valid for bigger parts;
- Latest results presented in the literature [38,39] have indicated, by means of a cytotoxicity test, that heating the Ti6Al4V alloy during the SPIF process does not affect its biocompatibility. Corroborated by the superior plastic behavior of the heated material, these results narrow the application range of SPIF at room temperature. However, there are still reasons favoring the approach of SPIF at room temperature, from the points of view of roughness, costs (related to equipment complexity and energy consumption), and degree of control.

4. Conclusions

After conducting the experimental program, from a technological point of view it can be concluded that Ti6Al4V titanium alloy may be used for manufacturing cranioplasty plates, by processing by means of SPIF at room temperature, if the following technological aspects are considered:

- To reduce friction, the punch must rotate around its axis. A rotation speed between 150 and 300 rev/min was found during this experimental work to be appropriate;
- Theoretically, the working feed does not influence the formability of the part; however, it was found that working feeds greater than 200 mm/min may lead to crack occurrence. However, working feed affects the productivity, which is not critically important for this kind of part. Cranioplasty plates are manufactured as prototypes, productivity having less importance from this point of view;
- The vertical step of the punch (no matter if circular or spiral toolpaths are used) must be smaller than 1 mm. Best results were achieved by using vertical steps of 0.6, 0.4, and 0.2 mm. A direct link was noticed between the vertical step and the maximum inclination angle ($\alpha°$) of the wall: the smaller the vertical step, the larger the achievable angle;
- Continuous toolpaths (which do not use lead-in/lead out entry/exit points) are the best approach, because entry/exit points may become stress concentrators leading to crack occurrence. Spatial spiral trajectories provide the best results, but for irregular shapes, generating them without the aid of CAM software is difficult. Another solution is to use contour curves spaced on the Z axis as

trajectories (circles or another curves) and distribute the lead-in/lead out points on the surface of the part (avoiding placing them close to each other);

- The maximum achievable wall angle was found $\alpha = 38°$;
- The best results, from the point of view of both the accuracy and surface roughness, were obtained using the punch with diameter $d_p = 10$ mm (compared with the punch with $d_p = 8$ mm);
- The accuracy of the wall angle was not significantly influenced by the diameter of the punch or by the vertical step. Also, the toolpaths did not influence it. The explanation for this may be found in the fact that the low plasticity of the Ti6Al4V titanium alloy does not lead to significant values of the springback;
- The roughness of the parts was influenced by the vertical step directly, a decrease in the vertical step leading to a decrease in the roughness value.
- Further research will be oriented in the following directions:
- The influence of rotational speed and working feed upon the accuracy of the part will be studied in more detail;
- For the time being, the overall dimensions of the parts were limited by the size of the experimental layout (mainly the size of the active plate and the working space of the CNC machine-tool). A new layout will be designed and implemented to test how greater overall dimensions of the part influence its manufacturability by means of ASPIF at room temperature;
- Industrial robots will be used as technological equipment to test if superior kinematics (more complex processing trajectories) can improve the manufacturability of the parts.

Author Contributions: S.G.R., M.T. and O.B. designed and set up the experimental layouts and the experimental program, R.E.B. set up the machining strategies, performed the AHP analysis, and wrote the paper, S.G.R., C.G., C.B. and A.L.C. performed the formability measurements and data analysis.

Funding: This research was funded by the Romanian Ministry of Research and Innovation CCCDI-UEFISCDI, project number PN-III-P1-1.2-PCCDI-2017-0446/nr. 82PCCDI/2018, within PNCDI III, project title: "Intelligent manufacturing technologies for advanced production of parts".

Conflicts of Interest: The authors declare no conflict of interest.

References

1. Rack, H.J.; Qazi, J.I. Titanium alloys for biomedical applications. *Mater. Sci. Eng. C* **2006**, *26*, 1269–1277. [CrossRef]
2. Geetha, M.; Singh, A.K.; Asokamani, R.; Gogia, A.K. Ti based biomaterials, the ultimate choice for orthopaedic implants—A review. *Prog. Mater. Sci.* **2009**, *54*, 397–425. [CrossRef]
3. Williams, D.F. On the mechanisms of biocompatibility. *Biomaterials* **2008**, *29*, 2941–2953. [CrossRef] [PubMed]
4. Aydin, S.; Kucukyuruk, B.; Abuzayed, B.; Aydin, S.; Sanus, G.Z. Cranioplasty: Review of materials and techniques. *J. Neurosci. Rural Pract.* **2011**, *2*, 162–167. [PubMed]
5. Servadei, F.; Iaccarino, C. The Therapeutic cranioplasty still needs an ideal material and surgical timing. *World Neurosurg.* **2015**, *83*, 133–135. [CrossRef] [PubMed]
6. Joffe, J.M.; Nicoll, S.R.; Richards, R.; Linney, A.D.; Harris, M. Validation of computer-assisted manufacture of titanium plates for cranioplasty. *Int. J. Oral Maxillofac. Surg.* **1999**, *28*, 309–313. [CrossRef]
7. Wiggins, A.; Austerberry, R.; Morrison, D.; Kwok, H.M.; Honeybul, S. Cranioplasty with custom-made titanium plates—14 years experience. *Neurosurgery* **2013**, *72*, 248–256. [CrossRef] [PubMed]
8. Heissler, E.; Fischer, F.-S.; Boiouri, S.; Lehrnann, T.; Mathar, W.; Gebhardt, A.; Lanksch, W.; Bler, J. Custom-made cast titanium implants produced with CAD/CAM for the reconstruction of cranium defects. *Int. J. Oral Maxillofac. Surg.* **1998**, *27*, 334–338. [CrossRef]
9. Cabraja, M.; Klein, M.; Lehmann, T.-N. Long-term results following titanium cranioplasty of large skull defects. *Neurosurg. Focus* **2009**, *26*, 1–7. [CrossRef] [PubMed]
10. Bhargava, D.; Bartlett, P.; Russell, J.; Liddington, M.; Tyagi, A.; Chumas, P. Construction of titanium cranioplasty plate using craniectomy bone flap as template. *Acta Neurochir.* **2010**, *152*, 173–176. [CrossRef] [PubMed]

11. Chen, J.-J.; Liu, W.; Li, M.-Z.; Wang, C.-T. Digital manufacture of titanium prosthesis for cranioplasty. *Int. J. Adv. Manuf. Technol.* **2006**, *27*, 1148–1152. [CrossRef]

12. Che-Haron, C.H.; Jawaid, A. The effect of machining on surface integrity of titanium alloy Ti–6% Al–4% V. *J. Mater. Process. Technol.* **2005**, *166*, 188–192. [CrossRef]

13. Jeswiet, J.; Micari, F.; Hirt, G.; Bramley, A.; Duflou, J.; Allwood, J. Asymmetric single point incremental forming of sheet metal. *CIRP Ann. Manuf. Technol.* **2005**, *54*, 623–650. [CrossRef]

14. Behera, A.K.; de Sousa, R.A.; Ingarao, G.; Oleksik, V. Single point incremental forming: An assessment of the progress and technology trends from 2005 to 2015. *J. Manuf. Process.* **2017**, *27*, 37–62. [CrossRef]

15. Gatea, S.; Ou, H.; McCartney, G. Review on the influence of process parameters in incremental sheet forming. *Int. J. Adv. Manuf. Technol.* **2016**, *87*, 479–499. [CrossRef]

16. Hussain, G.; Gao, L.; Hayat, N.; Cui, Z.; Pang, Y.C.; Dar, N.U. Tool and lubrication for negative incremental forming of a commercially pure titanium sheet. *J. Mater. Process. Technol.* **2008**, *203*, 193–201. [CrossRef]

17. Hamilton, K.; Jeswiet, J. Single point incremental forming at high feed rates and rotational speeds: Surface and structural consequences. *CIRP Ann. Manuf. Technol.* **2010**, *59*, 311–314. [CrossRef]

18. Kim, Y.H.; Park, J.J. Effect of process parameters on formability in incremental forming of sheet metal. *J. Mater. Process. Technol.* **2002**, *130–131*, 42–46. [CrossRef]

19. Durante, M.; Formisano, A.; Langella, A.; Memola Capece Minutolo, F. The influence of tool rotation on an incremental forming process. *J. Mater. Process. Technol.* **2009**, *209*, 4621–4626. [CrossRef]

20. Ambrogio, G.; Gagliardi, F.; Bruschi, S.; Filice, L. On the high-speed Single Point Incremental Forming of titanium alloys. *CIRP Ann. Manuf. Technol.* **2013**, *62*, 243–246. [CrossRef]

21. Fan, G.; Gao, L.; Hussain, G.; Wu, Z. Electric hot incremental forming: A. novel technique. *Int. J. Mach. Tools Manuf.* **2008**, *48*, 1688–1692. [CrossRef]

22. Ambrogio, G.; Filice, L.; Gagliardi, F. Formability of lightweight alloys by hot incremental sheet forming. *Mater. Des.* **2012**, *34*, 501–508. [CrossRef]

23. Palumbo, G.; Brandizzi, M. Experimental investigations on the single point incremental forming of a titanium alloy component combining static heating with high tool rotation speed. *Mater. Des.* **2012**, *40*, 43–51. [CrossRef]

24. Göttmann, A.; Diettrich, J.; Bergweiler, G.; Bambach, M.; Hirt, G.; Loosen, P.; Poprawe, R. Laser-assisted asymmetric incremental sheet forming of titanium sheet metal parts. *Prod. Eng.* **2011**, *5*, 263–271. [CrossRef]

25. Xu, D.; Wu, W.; Malhotra, R.; Chen, J.; Lu, B.; Cao, J. Mechanism investigation for the influence of tool rotation and laser surface texturing (LST) on formability in single point incremental forming. *Int. J. Mach. Tools Manuf.* **2013**, *73*, 37–46. [CrossRef]

26. Xu, D.K.; Lu, B.; Cao, T.T.; Zhang, H.; Chen, J.; Long, H.; Cao, J. Enhancement of process capabilities in electrically-assisted double sided incremental forming. *Mater. Des.* **2016**, *92*, 268–280. [CrossRef]

27. Feng, B.; Chen, J.Y.; Oi, S.K.; He, L.; Zhao, J.Z.; Zhang, X.D. Characterization of surface oxide films on titanium and bioactivity. *J. Mater. Sci. Mater. Med.* **2002**, *13*, 457–464.

28. Guleryuz, H.; Cimenoglu, H. Surface modification of a Ti–6Al–4V alloy by thermal oxidation. *Surf. Coat. Technol.* **2005**, *192*, 164–170. [CrossRef]

29. Duarte, L.T.; Bolfarini, C.; Biaggio, S.R.; Rocha-Filho, R.C.; Nascente, P.A.P. Growth of aluminum-free porous oxide layers on titanium and its alloys Ti–6Al–4V and Ti–6Al–7Nb by micro-arc oxidation. *Mater. Sci. Eng. C* **2014**, *41*, 343–348. [CrossRef] [PubMed]

30. Guleryuz, H.; Cimenoglu, H. Oxidation of Ti–6Al–4V alloy. *J. Alloys Compd.* **2009**, *472*, 241–246. [CrossRef]

31. Chen, M.; Li, W.; Shen, M.; Zhu, S.; Wang, F. Glass–ceramic coatings on titanium alloys for high temperature oxidation protection: Oxidation kinetics and microstructure. *Corros. Sci.* **2013**, *74*, 178–186. [CrossRef]

32. Zhang, Y.; Maa, G.-R.; Zhang, X.-C.; Li, S.; Tu, S.-T. Thermal oxidation of Ti-6Al–4V alloy and pure titanium under external bending strain: Experiment and modelling. *Corros. Sci.* **2017**, *122*, 61–73. [CrossRef]

33. Du, H.L.; Datta, P.K.; Lewis, D.B.; Burnell-Gray, J.S. Air oxidation behavior ofTi-6Al-4 V alloy between 650 and 850 °C. *Corros. Sci.* **1994**, *36*, 631–642. [CrossRef]

34. McLachlan, D.R.C. Aluminium and the risk for Alzheimer's disease. *Environmetrics* **1995**, *6*, 233–275. [CrossRef]

35. Aniołek, K.; Kupka, M.; Barylski, A.; Dercz, G. Mechanical and tribological properties of oxide layers obtained on titanium in the thermal oxidation process. *Appl. Surf. Sci.* **2015**, *357*, 1419–1426. [CrossRef]

36. Luo, Y.; Chen, W.; Tian, M.; Teng, S. Thermal oxidation of Ti6Al4V alloy and its biotribological properties under serum lubrication. *Tribol. Int.* **2015**, *89*, 67–71. [CrossRef]

37. Aniołek, K. The influence of thermal oxidation parameters on the growth of oxide layers on titanium. *Vacuum* **2017**, *144*, 94–100. [CrossRef]

38. Ambrogio, G.; Sgambitterra, E.; De Napoli, L.; Gagliardi, F.; Fragomeni, G.; Piccininni, A.; Gugleilmi, P.; Palumbo, G.; Sorgente, D.; La Barbera, L.; et al. Performances analysis of titanium prostheses manufactured by superplastic forming and incremental forming. *Procedia Eng.* **2017**, *183*, 168–173. [CrossRef]

39. Palumbo, G.; Sorgente, D.; Vedani, M.; Mostaed, E.; Hamidi, M.; Gastaldi, D.; Villa, T. Effects of superplastic forming on modification of surface properties of Ti alloys for biomedical applications. *J. Manuf. Sci. Eng.* **2018**, *140*, 10. [CrossRef]

40. Ingarao, G.; Ambrogio, G.; Gagliardi, F.; Di Lorenzo, R. A sustainability point of view on sheet metal forming operations: Material wasting and energy consumption in incremental forming and stamping processes. *J. Clean. Prod.* **2012**, *29–30*, 255–268. [CrossRef]

41. Bagudanch, I.; Garcia-Romeu, M.L.; Ferrer, I.; Lupiañez, J. The effect of process parameters on the energy consumption in single point incremental forming. *Procedia Eng.* **2013**, *63*, 346–353. [CrossRef]

42. Ingarao, G.; Vanhove, H.; Kellens, K.; Duflou, J.R. A comprehensive analysis of electric energy consumption of single point incremental forming processes. *J. Clean. Prod.* **2014**, *67*, 173–186. [CrossRef]

43. Ambrogio, G.; Ingarao, G.; Gagliardi, F.; Di Lorenzo, R. Analysis of energy efficiency of different setups able to perform single point incremental forming (SPIF) Processes. *Procedia CIRP* **2014**, *15*, 111–116. [CrossRef]

44. Breaz, R.; Bologa, O.; Tera, M.; Racz, G. Researches Regarding the Use of Complex Trajectories and Two Stages Processing in Single Point Incremental Forming of Two Layers Sheet. *Circles* **2012**, *91*, 128.

45. Breaz, R.; Bologa, O.; Tera, M.; Racz, G. Computer assisted techniques for the incremental forming technology. In Proceedings of the IEEE 18th Conference on Emerging Technologies & Factory Automation (ETFA), Cagliari, Italy, 10–13 September 2013.

46. Cotigă, C.; Bologa, O.; Racz, S.G.; Breaz, R.E. Researches regarding the usage of titanium alloys in cranial implants. *Appl. Mech. Mater.* **2014**, *657*, 173–177. [CrossRef]

47. Kim, B.J.; Hong, K.S.; Park, K.J.; Park, D.H.; Chung, Y.G.; Kang, S.H. Customized cranioplasty implants using three-dimensional printers and polymethyl-methacrylate casting. *J. Korean Neurosurg. Soc.* **2012**, *52*, 541–546. [CrossRef] [PubMed]

48. Digital Evolution of Cranial Surgery. Available online: http://www.renishaw.com/en/digital-evolution-of-cranial-surgery--38602 (accessed on 7 August 2018).

49. Saaty, T.L. *The Analytic Hierarchy Process: Planning, Priority Setting, Resource Allocation*; McGraw-Hill: New York, NY, USA, 1980; p. 287.

50. Saaty, T.L. *Decision Making for Leaders: The Analytic Hierarchy Process for Decisions in a Complex Word*; RWS Publication: Pittsburgh, PA, USA, 1990.

51. Alonso, J.; Lamata, T.M. Consistency in the analytic hierarchy process: A new approach. *Int. J. Uncertain. Fuzziness Knowl. Based Syst.* **2006**, *14*, 445–459. [CrossRef]

52. Cabala, P. Using the analytic hierarchy process in evaluating decision alternatives. *Oper. Res. Decis.* **2010**, *20*, 5–23.

53. Hurley, W.J. The analytic hierarchy process: A note on an approach to sensitivity which preserves rank order. *Comput. Oper. Res.* **2001**, *28*, 185–188. [CrossRef]

54. Williams, L.R.; Fan, K.F.; Bentley, R.P. Custom-made titanium cranioplasty: Early and late complications of 151 cranioplasties and review of the literature. *Int. J. Oral Maxillofac. Surg.* **2015**, *44*, 599–608. [CrossRef] [PubMed]

55. Williams, L.; Fan, K.; Bentley, R. Titanium cranioplasty in children and adolescents. *J. Cranio Maxillofac. Surg.* **2016**, *44*, 789–794. [CrossRef] [PubMed]

Article

Implant Treatment in Atrophic Maxilla by Titanium Hybrid-Plates: A Finite Element Study to Evaluate the Biomechanical Behavior of Plates

María Prados-Privado [1,2,*], Henri Diederich [3] and Juan Carlos Prados-Frutos [4]

1   Department Continuum Mechanics and Structural Analysis, Higher Polytechnic School, Carlos III University, Avenida de la Universidad, 30, 28911 Leganés, Madrid, Spain
2   Research Department, ASISA Dental, Calle José Abascal, 32, 28003 Madrid, Spain
3   Private Practice, 51 Avenue Pasteur, L2311 Luxembourg, Luxembourg; hdidi@pt.lu
4   Department of Medicine and Surgery, Faculty of Health Sciences, Rey Juan Carlos University, Avenida de Atenas s/n, 28922 Alcorcón, Madrid, Spain; juancarlos.prados@urjc.es
*   Correspondence: mprados@ing.uc3m.es; Tel.: +34-914-443-012

Received: 31 May 2018; Accepted: 24 July 2018; Published: 25 July 2018

Abstract: A severely atrophied maxilla presents serious limitations for rehabilitation with osseointegrated implants. This study evaluated the biomechanical and long-term behavior of titanium hybrid-plates in atrophic maxilla rehabilitation with finite elements and probabilistic methodology. A three-dimensional finite element model based on a real clinical case was built to simulate an entirely edentulous maxilla with four plates. Each plate was deformed to become accustomed to the maxilla's curvature. An axial force of 100 N was applied in the area where the prosthesis was adjusted in each plate. The von Mises stresses were obtained on the plates and principal stresses on maxilla. The difference in stress between the right and left HENGG-1 plates was 3%, while between the two HENGG-2 plates it was 2%, where HENGG means Highly Efficient No Graft Gear. A mean maximum value of 80 MPa in the plates' region was obtained, which is a lower value than bone resorption stress. A probability cumulative function was computed. Mean fatigue life was 1,819,235 cycles. According to the results of this study, it was possible to conclude that this technique based on titanium hybrid-plates can be considered a viable alternative for atrophic maxilla rehabilitation, although more studies are necessary to corroborate the clinical results.

Keywords: atrophic maxilla; titanium hybrid-plates; finite element analysis; biomechanical analysis

---

1. Introduction

The reconstruction of an atrophic maxilla has always been a challenge [1] because of anatomical and clinical factors due to the serious limitations for conventional implant placement [2]. These limitations are related to the amount of bone, which remains insufficient for the conventional placement of a dental implant [3]. The maxillary bone volume has been classified, among other authors, by Cawood and Howel in five grades (I to V). Grades IV and V are considered as extreme atrophies [4]. The most common alternatives in atrophic maxilla rehabilitation are bone grafting [5], pterygoid [6] or zygomatic implants [7], bone regeneration (with or without mesh) [8,9], and finally, short implants [10].

Bone grafting is the most common technique in the reconstruction of an atrophic maxilla [2]. The goal of hard tissue augmentation is to provide an adequate bone volume for ideal implant placement and to support soft tissue for optimal esthetics and function. Zygomatic implants present a viable alternative for this kind of treatment given their design with self-tapping screw and the appropriate length as this kind of implant can be placed in the bone with very good quality and excellent mechanical behavior [2,7]. Pterygoid implants have the advantage of allowing anchorage in the pterygomaxillary region, eliminating the need for sinus lifts or bone grafts. Additionally, pterygoid implants can eliminate posterior cantilever and improve axial loading [2,11]. Finally, short implants are widely used and have demonstrated their efficiency on implant treatment in atrophic jaw and maxilla [12,13].

The first hybrid plates were introduced by G. Scortecci in July 2000 and were first used in the same year in a patient with a fractured atrophic mandible. They had a large base plate (25, 33, or 43 mm long, 7, 9, or 12 mm wide) [14]. These kinds of plates can be adjusted to the maxilla curvature and put in the best place to maintain occlusal function [15].

There are technical differences between the hybrid plates employed and the sub-periosteal implants. The first difference is that the hybrid plates are made of titanium grade II and machined from a block, and the sub-periosteal implants are made of chrome cobalt and cast individually. The second difference is that hybrid plates are flexible and can be adjusted in situ while sub-periosteal implants are rigid and modeled using an initial impression of the bone site. Sub-periosteal implants can only be cemented to the prosthesis while hybrid plates have a screw connection. To use them, it is necessary to make a groove in the bone to receive the plates so that they may become osseointegrated, which are then fixed with osteosynthesis screws.

Hybrid plates are made of titanium grade II, which is the main cp Ti used for industrial dental implant applications [16]. It is recognized that titanium and its alloys are biomaterials with the best in vivo behavior [17]. Due to the excellent biocompatibility of this material, it is very common to use it for biomedical applications such as dental implants ad hybrid-plates [18]. This biocompatibility is provided by the following properties of titanium: low level of conductivity, high corrosion resistance, thermodynamic state at physiological pH values, and low ion-formation tendency in aqueous environments [19,20].

In that sense, the main differences between hybrid plates and other alternatives to treat atrophic maxilla as pterygoid, zygomatic, or short implants are a titanium alloy. Implants are mostly manufactured by titanium alloy Ti6Al4V.

The protocol employed in this study is called Cortically Fixed @ Once (CF@O). It is an alternative to conventional implant placement for severe atrophied maxilla and mandible. This technique uses plates and pterygoid implants. Plates are fixed to the bone with osteosynthesis screws. There are four types of plates available that differ in size and morphology, although in this study, only two of them were used. This technique has its origins in basal implantology, which was developed by Dr. Scortecci in the early 1980s when he proposed the Diskimplant®, a disc implant system that was inserted laterally, and which he refined over the next few years [21]. Several basal implants were developed during the 1980s and the 2000s with different geometric forms and with perforations over the surface to improve the blood supply around the implant [14].

In the last few decades, the finite element method (FEM) has become very popular in the field of biomechanics as it is a useful tool to numerically calculate aspects such as stresses and strains, and to evaluate the mechanical behavior of biomaterials and human tissues, considering the difficulty in making such an assessment in vivo [22,23].

Dental implants and their components including hybrid-plates are subjected to cyclic loads. Therefore, fatigue of materials is introduced in all dental rehabilitations. Results with a good accuracy are essential in dental studies as fatigue is very sensitive to many parameters [22].

Owing to the variety of techniques for atrophic maxillary implant rehabilitation, a new technique based on innovative hybrid titanium plates is described and numerically analyzed. The aim of this study was to evaluate the biomechanical and long-term behavior of CF@O plates on a completely edentulous and atrophic maxilla by employing a finite element analysis and a probabilistic fatigue approach.

2. Materials and Methods

2.1. Description of the Protocol

The CF@O protocol is an alternative to the existing treatment of the atrophied maxilla and mandible. This technique is less invasive than the conventional procedures, and implants can be loaded with a definitive restoration after 6–10 days. Between the surgery and the definitive prosthesis, the patient has a provisional prosthesis, which is a fixed immediate loading prosthesis made of resin installed on plates.

This protocol does not need a sinus lift for the rehabilitation in cases of atrophied maxilla nor bone graft in maxilla and mandible and is based on both traditional implantology methods, combined with the most modern tools.

The Cortically Fixed @ Once protocol is applicable to edentulous maxillae and mandibles, and to unilateral and bilateral edentulism in maxilla and mandible. This protocol is indicated in the following assumptions [24]:

- Reduced bone volume in the upper and lower jaws such as stage D or E according to the Lekholm and Zarb classification of bone quality.
- Severely reduced vertical bone height over the trajectory of the mandibular canal in the lower jaw where there is insufficient vertical bone height to place conventional implants.
- Where short implants are not deemed justified, or in cases where bone reconstruction is not feasible.
- In very sharp dentoalveolar ridges.

This protocol can be used in patients aged between 35 and 90 years and is performed under local anesthesia. It can be done in nearly all cases of atrophy except in the case of egg shell everywhere in the maxilla and 10 mm of residential bone in the mandible. A stereolithographic model based on CT scan is made. After surgery, a temporary fixed bridge is employed. Amoxicillin is prescribed (2 g a day for 10 days) and in case of pain, ibuprofen 600 (1–3 g a day). Finally, there are no restrictions on food after one month after surgery.

The treatment plan starts with signed patient consent. Then, an open flap in the maxilla, as in this study, is made from the left tuberosity along the crest till the canine region. Two hybrid plates HENGG-1 and HENGG-2 are fixed with osteosynthesis screws and covered. The flap is then closed on the left and right with polytetrafluoroethylene polymer (PTFE) monofilament nonabsorbable suture.

Depending on the atrophy, three types of prosthesis are available: a metal acrylic for a big atrophy and a metal ceramic or zirconia when there is less atrophy with enough bone. Another point to consider is the contribution of the pterygoid implants, which contribute to prosthesis fixation.

From 2013, 155 patients between the ages of 38 and 85 (95 were female and 60 were male and 105 were in the maxilla and 45 in the mandible) were treated with the protocol detailed previously, resulting in three failures of plates in the maxilla and two in the jaw. These lost treatments were related to infection processes in the soft tissues. After a follow-up period of one year, there was a clinical and radiographical check to confirm that the plates were fixed and without complication.

The finite element analysis employed in this study simulated a real case of a 58-year-old female, with patient consistent, who wanted fixed teeth in the maxilla in a compromised bone. The treatment consisted of two pterygoid implants, four hybrid plates fixed with osteosynthesis screws, and a metal acrylic bridge ten days later [25]. In this instance, only the plates were analyzed.

2.2. Plates

The plates used in the CF@O protocol are very thin, lightweight, and highly flexible, and therefore may be adapted to any bone anatomy. In this study, the two plates employed are detailed in Figure 1:

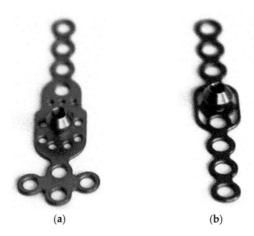

Figure 1. Plates employed in this study: (**a**) HENGG-1; (**b**) HENGG-2.

The HENGG-1 plate is appropriated for atrophied maxilla and is fixed with the zygomatic bone and the palate. The HENGG-2 plate is recommended for premaxilla, and the retromolar region in case of pencil mandible.

The plates are milled in a single piece and may be tilted in two axes to ensure that the implant fits the bone perfectly by manual shaping, making them isoelastic and able to mimic bone. They are minimally invasive and totally adjustable; they can be tilted at 90 degrees and the number of vents needed can be reduced as required, depending on the bone available at the site. They can also be twisted to fit the mandibular anatomy. They are stabilized and fixed by osteosynthesis screws, which give a strong cortical anchorage.

2.3. Finite Element Reconstruction

Geometric characteristics of the plates employed in the present study are shown in Figures 2 and 3.

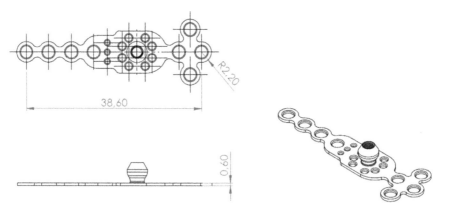

Figure 2. Plate HENGG-1.

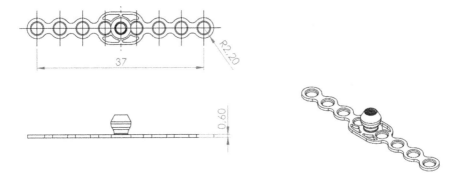

Figure 3. Plate HENGG-2.

All three-dimensional plates were adjusted to the anatomic characteristics of the maxilla (Figure 4b) as the common procedure in a real case (Figure 4a).

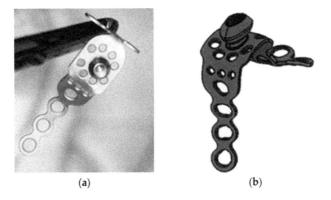

(a)   (b)

Figure 4. Plate adjusted to the anatomic characteristics of the maxilla: (**a**) Plate deformed before being placed in the patient; (**b**) three-dimensional model of a plate deformed.

The finite element model reproduced the case detailed previously, which is represented in Figure 5. Geometry of the maxilla was obtained using CT and transformed to the STL format. Slice increment was 0.5 mm, according to other studies in the literature. All data in DICOM format were imported into the software package Mimics 10.0 (Materialize, Leuven, Belgium) for the construction of the 3-D model. Plate HENGG-1 was placed in the molar region and HENGG-2 in the premaxilla.

Finally, Figure 6 represents the three-dimensional finite element assembly employed to reproduce the clinical case detailed in this study.

The maxilla in STL format was imported into SolidWorks 2016 (Dassault Systèmes, SolidWorks Corp., Concord, MA, USA) where the assembly with the four plates was done (Figure 6). The trabecular bone was 1 mm thick [26].

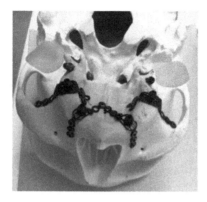

Figure 5. Model employed to reproduce in the finite element model.

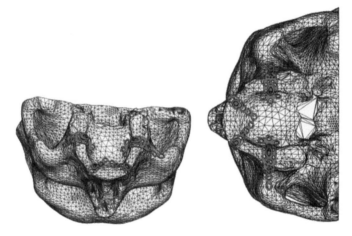

Figure 6. Finite element model employed in this study.

2.4. Material Properties

All materials were considered isotropic, linearly elastic, and homogeneous. The properties of the materials are detailed in Table 1.

Table 1. Material properties employed in this model.

Title	Young's Modulus	Poisson's Ratio
Plates (titanium grade II) [27]	105 GPa	0.37
Maxilla: cortical bone [26]	13.7 GPa	0.3
Maxilla: trabecular bone [26]	1.37 GPa	0.3

2.5. Meshing

Mesh generation was done in SolidWorks 2016 (Dassault Systèmes, SolidWorks Corp., Concord, MA, USA). All components were meshed with a fine mesh and all regions of stress concentration that were of interest were manually refined. The three-dimensional model presented a total of 432,404 nodes and 294,104 elements. The convergence criterion was a change of less than 5% in the von Mises stress in the model [28] (Figure 7).

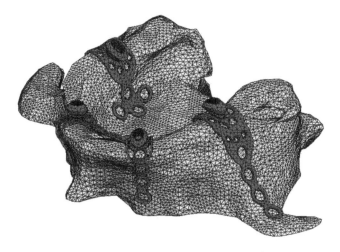

Figure 7. Finite element mesh (isometric view).

2.6. Boundary Conditions and Loading Configuration

The model was subjected to a rigid fixation restriction in the upper and lateral maxilla to prevent displacement in the *x*, *y*, and *z* axes (Figure 8). Plates were in contact with the maxilla and a nonpenetration condition was also added to prevent interferences during the execution process between the plates and the maxilla.

A load of 100 N [29,30] was directly applied perpendicular to the area where the prosthesis was fixed to the plate as detailed in Figure 8.

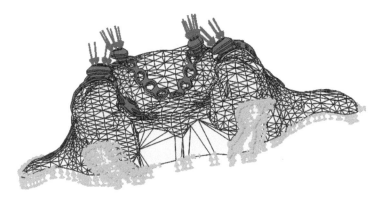

Figure 8. Three-dimensional posterior view of maxilla with load application and boundary conditions. Loads are represented in blue arrows and the rigid fixation restriction is represented in green.

2.7. Probabilistic Fatigue Model

In addition to the previous deterministic finite element analysis, a probabilistic fatigue model at the crack nucleation stage was also implemented. This stage is the most important regarding dental components and, hybrid plates, life, in particular [31]. As Riahi et al. detailed in their study, the probabilistic finite element method is a viable tool to estimate the influence of the stochastic properties of loads, material properties, and geometry on the response [32]. The methodology employed in this study was based on a cumulative damage B-model, which is constructed from the Probabilistic Finite Element Method (PFEM) results and computed for every random variable here considered: the Young's modulus (105 ± 10 GPa) and the applied loads (100 ± 10 N) [31]. The input random variables considered in this study were handled via its first order Taylor series expansion. Once all the sensitivities of the random variables are known, it is possible to apply the mean and variance operator. All the sensitivities of the random fields involved, such as displacements field, strain field, and stress field can be obtained.

Bogdanoff and Kozin (B-K) created a number of probabilistic models of cumulative damage based on ideas from Markov chains. This study employed one they called the B-model of unit steps, for its simplicity and suitability to the physical description of the process of fatigue in the crack initiation stage. The hypotheses that serve as a basis for the expansion of the B-K unit step model are [33]:

(1) Damage cycles (DC) are repetitive and of constant severity.
(2) The levels of damage a component will go through until final failure are discrete $(1, 2, \ldots, j, \ldots, b)$, and failure occurs at the last level of damage (b). This hypothesis merely discretizes the total life of the component in b levels.
(3) The accumulation of damage that occurs in each DC depends only on the DC itself and the level of damage of the component at the start of said DC.
(4) The level of damage in each DC can only be increased from the occupied level at the beginning of said DC to the next immediate level.

As damage cycles have been defined as constant severity, the Probability Transition Matrix (P) will be unique and expresses the probability that each DC must be in the same level or the probability will jump to the next DC. This matrix depends on the p_j (probability of remaining in the same DC) and q_j (probability of jumping to the next DC) and is detailed in Equation (1).

$$P = \begin{pmatrix} p_1 & q_1 & 0 & \cdots & 0 & 0 \\ 0 & p_2 & q_2 & 0 & \cdots & 0 \\ 0 & 0 & p_3 & q_3 & \cdots & 0 \\ \vdots & \vdots & \vdots & \vdots & \ddots & \vdots \\ 0 & 0 & 0 & \cdots & p_{b-1} & q_{b-1} \\ 0 & 0 & 0 & \cdots & 0 & 1 \end{pmatrix} \tag{1}$$

The new vector p_x is a vector showing the distribution of damage levels for time $t = x$. Using the results of Markov chains, vector p_x is:

$$p_x = p_{x-1}P = p_0 P^x \text{ with } x = 0, 1, 2 \ldots . \tag{2}$$

Finally, to compute the fatigue life estimators, Neuber's rule and a random formulation of the Coffin and Basquin–Manson expressions were employed. Neuber's rule relates the levels of elastic stress and strain obtained by a linear elastic analysis with actual levels of stress and strain, in accordance to the elastic-plastic behavior material [34].

Coffin, for the elastic component of deformations, and Basquin and Manson, for the elastic-plastic component, proposed a nonexplicit relationship between the fatigue life cycles in the nucleation stage of a component and the amplitudes of strain. This relation is shown in Equation (3)

$$\frac{\Delta \varepsilon_{ep}}{2} = \frac{\sigma'_f}{E} \left(2N_f \right)^b + \varepsilon'_f \left(2N_f \right)^c \tag{3}$$

where $\Delta\varepsilon_{ep}$ is the range of elastic-plastic strain suffered by the component at the crack initiation area; σ'_f is the fatigue resistance coefficient; ε'_f is the fatigue ductility coefficient; b is the fatigue resistance exponent; c is the fatigue ductility exponent; E is the modulus of elasticity; and N_f is the fatigue life cycles.

The materials properties necessary to solve this probabilistic model are detailed in Table 2:

Table 2. Material properties for the probabilistic model.

Parameter	Value	Parameter	Value
$\Delta\varepsilon_{ep}$	0.352×10^{-2}	σ'_f	0.0140×10^6
b	-0.1203	ε'_f	0.1701
c	-0.349		

3. Results

3.1. Plates

To obtain a correct clinical behavior, loads must be uniformly distributed throughout the four plates and transmit small stresses to the maxilla. Figure 9 shows the von Mises stress on the plates. Stress distribution along the plates is different because of the anatomical geometry of the maxilla, however, these differences on stress values are very small, as Figure 10 details.

von Mises (MPa)

195.32
179.03
162.89
146.58
130.24
113.91
97.65
81.38
65.10
48.83
32.55
16.28
9.37

Figure 9. The von Mises stress on plates in MPa.

In Figure 9, the maximum von Mises stresses appeared around the area where the prosthesis was adjusted to the plate and the body of the plate. The right HENGG-1 plate supported a stress of 185 MPa, while the same plate on the left had a maximum von Mises stress of 179 MPa. Stress on the right HENGG-2 plate was 168 MPa while in the plate placed on the left, it was 165 MPa.

According to Figure 10, the difference between the maximum von Mises stress in the HENGG-1 right and left plates was 3%, while the difference between the HENGG-2 right and left plates was 2%.

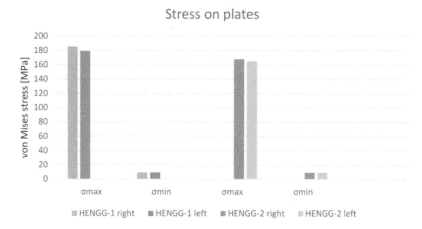

Figure 10. The von Mises stress values (in MPa).

3.2. Maxilla

All plates showed similar distribution patterns of maximum principal stress over the atrophic maxilla. The difference in the principal stress value between the four regions in contact with plates was 5%, with a mean maximum value of 80 MPa in the plates' region.

3.3. Long-Term Behavior

Failure probability of the situation detailed previously was obtained by employing the probabilistic methodology. The expected life computed was 1,819,235 ± 22.6 cycles. Then, the cumulative probability function was computed and represented in Figure 11.

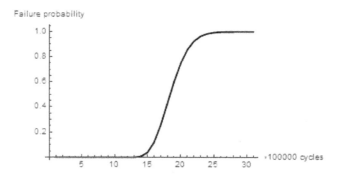

Figure 11. Cumulative probability function.

Figure 11 relates the probability of failure associated with each cycle load. As shown in the previous figure, the probability was equal to zero until 1,300,000 cycles.

4. Discussion

The biomechanical behavior of CF@O plates on a completely edentulous and atrophic maxilla was evaluated in the present study by employing finite element methods.

For a few years, there has been a trend towards minimally invasive implant treatment in very atrophic edentulous jaws and maxillae. The purpose of these concepts is to make an implant treatment with a shorter duration and smaller surgical risks [35]. The existence of insufficient bone can strongly influence the choice of the most appropriate rehabilitation in edentulous patients. There are several studies available in the literature that have employed different techniques to treat edentulous and atrophic maxilla, such as basal disk implants [36] or bone augmentation [37]. This study analyzed a new alternative based on titanium hybrid-plates. The accuracy of the results in numerical simulation studies depends on the precision of the model analyzed, the material properties, and the constraining conditions [38]. CT was used to model the geometry of the atrophic maxilla while the plates were provided by the manufacturer. A real clinical case with four titanium hybrid-plates in an atrophic maxilla was modeled and analyzed with the goal of knowing the biomechanical behavior of those plates.

This study had some assumptions and limitations. All materials were considered homogeneous, isotropic, and linearly elastic. Although these assumptions do not occur in clinical practice, they are common in finite element studies due to the challenges in establishing the properties of living tissues. These assumptions are consistent with other numerical studies [22,23,39]. In addition to these limitations, this work did not analyze the role of pterygoid implants as the goal was to study the biomechanical behavior of the plates. In that sense, ideal load distribution was considered.

The application of loads on the plates were supposed as an ideal transfer of loads from the prosthesis to that plate. If there is a good fit between the plate and the prosthesis, forces will be transmitted uniformly and as it was simulated and assumed in this study.

The ultimate strength in titanium grade II has been described as between 275 and 410 MPa and the ultimate tensile strength as 344 MPa. From these finite element analysis results, the maximum von Mises stress values in plates were lower than the ultimate strength [27]. The difference between the maximum von Mises stress between the distal plates was 3%, while the difference between the mesial plates was 2%. This difference was due to the different geometry of the maxilla in each area.

Küçükkurt et al. [40] compared the biomechanical behavior of different sinus floor elevations for dental implant placement. Under the condition of vertical loadings, von Mises stress in mesial implants were lower than our results in the plates in the case of lateral sinus lifting. However, the plates analyzed in this study obtained lower von Mises stress values than the prosthetic distal cantilever application and short implant placement. Regarding the distal implants, our plates obtained lower stresses than the prosthetic distal cantilever.

Ihde et al. [41] numerically analyzed baseplate implants with a vertical load of 114.6 N and a horizontal load 17.1 N and obtained a maximum von Mises stress of 400 MPa. Ihde et al. [42] detailed the von Mises stress in basal implants depending on the bone interface contact (BIC) degree. In this study, the maximum von Mises stress values were between 649 and 190 MPa. In both cases, the titanium hybrid plates used and analyzed in this study obtained lower stresses.

Kopp et al. [43] calculated the distribution of stress when basal implants in the mandible were loaded at two different stages of bone healing. They applied a load of 450 N located in the middle between the left molar and left canine implant and oriented in a vertical direction. Under these conditions, the von Mises stress in the basal implants was around 565 MPa.

The ultimate stress is an important value to understand the limits of the behavior of a material. According to physiological limits (ultimate stress), overloading in the cortical bone has been described as 170 MPa in compression and 100 MPa in tension [44]. Dos Santos et al. [45] detailed in their study that cortical bone resorption occurred when stress was higher than 167 MPa. Based on these limits, the values observed in this model were lower than those considered physiologic to bone tissue. As bone cannot be considered a ductile material, von Mises stress cannot be calculated in the maxilla. In this case, principal stresses have to be employed and calculated although some published studies have not used this kind of stress [39,45,46].

This is the reason why it is difficult to compare the results obtained in the atrophic maxilla of this study with the results provided in the bone in other published studies.

A good biomechanical behavior of plates is understood when a homogeneous stress is transferred to the bone. In this case, the maximum difference between the region of all four plates was 5%, meaning that the principal stress transferred from the plates to maxilla can be considered homogeneous.

Küçükkurt et al. [40] obtained a similar maximum principal stress in maxilla than our results in the case of short implant placement, and higher principal stress to our results in prosthetic distal cantilever application.

Clinical failures have generally been observed in the posterior maxilla region. Most of those failures were observed in bone types 3 and 4, with a highest probability of failures in bone type 4 [47]. According to the results obtained, the mean expected life in this case was 1,819,235 cycles. As Haug et al. detailed in their study [48], one year of in vivo service corresponds with, approximately, 200,000 cycles. Maló et al. obtained a satisfactory long-term outcome from patients with completely edentulous, severely atrophic maxillae supported by immediately loaded zygomatic implants alone, or in combination with conventional implants [49]. The same satisfactory results were detailed in Migliorança et al. who employed zygomatic implants placed lateral to the maxillary sinus and combined with conventional implants for a rehabilitation of the edentulous maxilla [50].

Further studies simulating these titanium hybrid-plates alternatives for atrophic maxilla and jaw that include dynamic forces that occur during chewing and consider the anisotropic and regenerative properties of bone are needed. Furthermore, some in vivo clinical trials are necessary to validate the model and to confirm the efficiency of this protocol. A numerical study of the combination of prosthesis-plates-implant under different functional conditions (bruxism and other parafunctions) just like the antagonist arcade is also necessary. Finally, a simulation of blood flow and bone regeneration around the plates is also necessary.

5. Conclusions

Based on the study results, it is possible to conclude that in terms of clinical application, the results indicate that titanium hybrid-plates proposed as an alternative to severe atrophic maxilla seems to have, from a mechanical point of view, a better behavior than conventional treatments such as prosthetic distal cantilever application and short implant placement. Titanium hybrid-plates distributed load to the maxilla with similar values as short implants but, with higher values than the prosthetic distal cantilever application. In any case, the resistance limits of bone and titanium were not exceeded. Long-term outcomes also seemed to be better than those clinical cases to treat atrophic maxilla. Finally, this technique can be considered as a viable alternative for atrophic maxilla rehabilitation, although more studies are necessary to corroborate the clinical results. As a clinical implication, this treatment allows the patient to be provided with functionally adequate prosthetic rehabilitation, which implies the recovery of their quality of life in a patient with a severe atrophy and, therefore, with an important challenge to a conventional implant treatment.

Author Contributions: M.P.-P. conceived, designed, and performed the analyses, evaluated the results, and wrote part of the paper. H.D. provided critical analysis. J.C.P.-F. provided critical analysis, interpretation of data, reviewed the literature, and wrote part of the paper.

Funding: This research was partially funded by Proclinic grant number A-285 and Instradent grant number A-274. Principal Researcher: Juan Carlos Prados-Frutos.

Conflicts of Interest: The authors declare no conflicts of interest.

References

1. Van der Mark, E.L.; Bierenbroodspot, F.; Baas, E.M.; de Lange, J. Reconstruction of an atrophic maxilla: Comparison of two methods. *Br. J. Oral Maxillofac. Surg.* **2011**, *49*, 198–202. [CrossRef] [PubMed]
2. Ali, S.; Karthigeyan, S.; Deivanai, M.; Kumar, A. Implant Rehabilitation For Atrophic Maxilla: A Review. *J. Indian Prosthodont. Soc.* **2014**, *13*, 196–207. [CrossRef] [PubMed]
3. Spencer, K. Implant based rehabilitation options for the atrophic edentulous jaw. *Aust. Dent. J.* **2018**, *63*, S100–S107. [CrossRef] [PubMed]

4. Cawood, J.I.; Howell, R.A. A classification of the edentulous jaws. *Int. J. Oral Maxillofac. Surg.* **1988**, *17*, 232–236. [CrossRef]

5. Chiapasco, M.; Zaniboni, M. Methods to Treat the Edentulous Posterior Maxilla: Implants With Sinus Grafting. *J. Oral Maxillofac. Surg.* **2009**, *67*, 867–871. [CrossRef] [PubMed]

6. Cucchi, A.; Vignudelli, E.; Franco, S.; Corinaldesi, G. Minimally Invasive Approach Based on Pterygoid and Short Implants for Rehabilitation of an Extremely Atrophic Maxilla. *Implant Dent.* **2017**, *26*, 639–644. [CrossRef] [PubMed]

7. Aparicio, C.; Manresa, C.; Francisco, K.; Claros, P.; Alández, J.; González-Martín, O.; Albrektsson, T. Zygomatic implants: Indications, techniques and outcomes, and the Zygomatic Success Code. *Periodontol. 2000* **2014**, *66*, 41–58. [CrossRef] [PubMed]

8. Gultekin, B.A.; Cansiz, E.; Borahan, M.O. Clinical and 3-Dimensional Radiographic Evaluation of Autogenous Iliac Block Bone Grafting and Guided Bone Regeneration in Patients With Atrophic Maxilla. *J. Oral Maxillofac. Surg.* **2017**, *75*, 709–722. [CrossRef] [PubMed]

9. Kaneko, T.; Nakamura, S.; Hino, S.; Norio, H.; Shimoyama, T. Continuous intra-sinus bone regeneration after nongrafted sinus lift with a PLLA mesh plate device and dental implant placement in an atrophic posterior maxilla: A case report. *Int. J. Implant Dent.* **2016**, *2*, 16. [CrossRef] [PubMed]

10. Alqutaibi, A.Y.; Altaib, F. Short dental implant is considered as a reliable treatment option for patients with atrophic posterior maxilla. *J. Evid. Based Dent. Pract.* **2016**, *16*, 173–175. [CrossRef] [PubMed]

11. Candel, E.; Peñarrocha, D.; Peñarrocha, M. Rehabilitation of the Atrophic Posterior Maxilla With Pterygoid Implants: A Review. *J. Oral Implantol.* **2012**, *38*, 461–466. [CrossRef] [PubMed]

12. Anitua, E.; Orive, G.; Aguirre, J.J.; Andía, I. Five-Year Clinical Evaluation of Short Dental Implants Placed in Posterior Areas: A Retrospective Study. *J. Periodontol.* **2008**, *79*, 42–48. [CrossRef] [PubMed]

13. Anitua, E.; Piñas, L.; Begoña, L.; Orive, G. Long-term retrospective evaluation of short implants in the posterior areas: Clinical results after 10–12 years. *J. Clin. Periodontol.* **2014**, *41*, 404–411. [CrossRef] [PubMed]

14. Scortecci, G.; Misch, C.; Benner, K. *Implants and Restorative Dentistry*; Martin Dunitz: London, UK, 2001.

15. Wu, C.; Lin, Y.; Liu, Y.; Lin, C. Biomechanical evaluation of a novel hybrid reconstruction plate for mandible segmental defects: A finite element analysis and fatigue testing. *J. Cranio-Maxillofac. Surg.* **2017**. [CrossRef] [PubMed]

16. Sidambe, A. Biocompatibility of Advanced Manufactured Titanium Implants—A Review. *Materials* **2014**, *7*, 8168–8188. [CrossRef] [PubMed]

17. Torres, Y.; Lascano, S.; Bris, J.; Pavón, J.; Rodriguez, J.A. Development of porous titanium for biomedical applications: A comparison between loose sintering and space-holder techniques. *Mater. Sci. Eng. C* **2014**, *37*, 148–155. [CrossRef] [PubMed]

18. Suzuki, K.; Takano, T.; Takemoto, S.; Ueda, T.; Yoshiari, M.; Sakurai, K. Influence of grade and surface topography of commercially pure titanium on fatigue properties. *Dent. Mater. J.* **2018**, *37*, 308–316. [CrossRef] [PubMed]

19. Elias, C.N.; Lima, J.H.C.; Valiev, R.; Meyers, M.A. Biomedical applications of titanium and its alloys. *JOM* **2008**, *60*, 46–49. [CrossRef]

20. Natali, A.N. *Dental Biomechanics*, 1st ed.; Taylor & Francis Group: London, UK, 2003; ISBN 0-415-30666-3.

21. Scortecci, G.; Bourbon, B. Dentures on the Diskimplant. *Rev. Fr. Prothes. Dent.* **1990**, *13*, 31–48.

22. Prados-Privado, M.; Bea, J.A.; Rojo, R.; Gehrke, S.A.; Calvo-Guirado, J.L.; Prados-Frutos, J.C. A New Model to Study Fatigue in Dental Implants Based on Probabilistic Finite Elements and Cumulative Damage Model. *Appl. Bionics Biomech.* **2017**, *2017*, 3726361. [CrossRef] [PubMed]

23. Ferreira, M.B.; Barão, V.A.; Faverani, L.P.; Hipólito, A.C.; Assunção, W.G. The role of superstructure material on the stress distribution in mandibular full-arch implant-supported fixed dentures. A CT-based 3D-FEA. *Mater. Sci. Eng. C* **2014**, *35*, 92–99. [CrossRef] [PubMed]

24. Agbaje, J.O.; Vrielinck, L.; Diederich, H. Rehabilitation of Edentulous Jaw Using Cortically Fixed at Once (Cf@O) Protocol: Proof of Principle. *Biomed. J. Sci. Tech. Res.* **2018**, *5*, 1–5. [CrossRef]

25. Diederich, H.; Junqueira, M.A.; Guimarães, S.L. Immediate Loading of an Atrophied Maxilla Using the Principles of Cortically Fixed Titanium Hybrid Plates. *Adv. Dent. Oral Health* **2017**, *3*, 1–3. [CrossRef]

26. Bhering, C.L.B.; Mesquita, M.F.; Kemmoku, D.T.; Noritomi, P.Y.; Consani, R.L.X.; Barão, V.A.R. Comparison between all-on-four and all-on-six treatment concepts and framework material on stress distribution in atrophic maxilla: A prototyping guided 3D-FEA study. *Mater. Sci. Eng. C* **2016**, *69*, 715–725. [CrossRef] [PubMed]

27. Boyer, R.; Welsch, G.; Collings, E.W. *Materials Properties Handbook: Titanium Alloys*; ASM International: Materials Park, OH, USA, 1994.

28. Peixoto, H.E.; Camati, P.R.; Faot, F.; Sotto-Maior, B.; Martinez, E.F.; Peruzzo, D.C. Rehabilitation of the atrophic mandible with short implants in different positions: A finite elements study. *Mater. Sci. Eng. C* **2017**, *80*, 122–128. [CrossRef] [PubMed]

29. Shimura, Y.; Sato, Y.; Kitagawa, N.; Omori, M. Biomechanical effects of offset placement of dental implants in the edentulous posterior mandible. *Int. J. Implant Dent.* **2016**, *2*, 17. [CrossRef] [PubMed]

30. Arat Bilhan, S.; Baykasoglu, C.; Bilhan, H.; Kutay, O.; Mugan, A. Effect of attachment types and number of implants supporting mandibular overdentures on stress distribution: A computed tomography-based 3D finite element analysis. *J. Biomech.* **2015**, *48*, 130–137. [CrossRef] [PubMed]

31. Prados-Privado, M.; Prados-Frutos, J.C.; Calvo-Guirado, J.L.; Bea, J.A. A random fatigue of mechanize titanium abutment studied with Markoff chain and stochastic finite element formulation. *Comput. Methods Biomech. Biomed. Eng.* **2016**, *19*, 1583–1591. [CrossRef] [PubMed]

32. Riahi, H.; Bressolette, P.; Chateauneuf, A. Random fatigue crack growth in mixed mode by stochastic collocation method. *Eng. Fract. Mech.* **2010**, *77*, 3292–3309. [CrossRef]

33. Bogdanoff, J.; Kozin, F. *Probabilistic Models of Cumulative Damage*; Wiley: New York, NY, USA, 1985.

34. Neuber, H. Theory of Stress Concentration for Shear-Strained Prismatical Bodies With Arbitrary Nonlinear Stress-Strain Law. *J. Appl. Mech.* **1961**, *28*, 544–550. [CrossRef]

35. Wentaschek, S.; Hartmann, S.; Walter, C.; Wagner, W. Six-implant-supported immediate fixed rehabilitation of atrophic edentulous maxillae with tilted distal implants. *Int. J. Implant Dent.* **2017**, *3*, 35. [CrossRef] [PubMed]

36. Odin, G.; Misch, C.E.; Binderman, I.; Scortecci, G. Fixed Rehabilitation of Severely Atrophic Jaws Using Immediately Loaded Basal Disk Implants After In Situ Bone Activation. *J. Oral Implantol.* **2012**, *38*, 611–616. [CrossRef] [PubMed]

37. Del Fabbro, M.; Rosano, G.; Taschieri, S. Implant survival rates after maxillary sinus augmentation. *Eur. J. Oral Sci.* **2008**, *116*, 497–506. [CrossRef] [PubMed]

38. Van Staden, R.C.; Guan, H.; Loo, Y.C. Application of the finite element method in dental implant research. *Comput. Methods Biomech. Biomed. Eng.* **2006**, *9*, 257–270. [CrossRef] [PubMed]

39. Almeida, E.O.; Rocha, E.P.; Júnior, A.C.F.; Anchieta, R.B.; Poveda, R.; Gupta, N.; Coelho, P.G. Tilted and Short Implants Supporting Fixed Prosthesis in an Atrophic Maxilla: A 3D-FEA Biomechanical Evaluation. *Clin. Implant Dent. Relat. Res.* **2015**, *17*, e332–e342. [CrossRef] [PubMed]

40. Küçükkurt, S.; Alpaslan, G.; Kurt, A. Biomechanical comparison of sinus floor elevation and alternative treatment methods for dental implant placement. *Comput. Methods Biomech. Biomed. Eng.* **2017**, *20*, 284–293. [CrossRef] [PubMed]

41. Ihde, S.; Goldmann, T.; Himmlova, L.; Aleksic, Z.; Kuzelka, J. Implementation of contact definitions calculated by FEA to describe the healing process of basal implants. *Biomed. Pap. Med. Fac. Univ. Palacky Univ. Olomouc* **2008**, *152*, 169–73. [CrossRef]

42. Ihde, S.; Goldmann, T.; Himmlova, L.; Aleksic, Z. The use of finite element analysis to model bone-implant contact with basal implants. *Oral Surg. Oral Med. Oral Pathol. Oral Radiol. Endodontol.* **2008**, *106*, 39–48. [CrossRef] [PubMed]

43. Kopp, S.; Kuzelka, J.; Goldmann, T.; Himmlova, L.; Ihde, S. Modeling of load transmission and distribution of deformation energy before and after healing of basal dental implants in the human mandible. *Biomed. Tech. Eng.* **2011**, *56*, 53–58. [CrossRef] [PubMed]

44. Pérez, M.A.; Prados-Frutos, J.C.; Bea, J.A.; Doblaré, M. Stress transfer properties of different commercial dental implants: A finite element study. *Comput. Methods Biomech. Biomed. Eng.* **2012**, *15*, 263–273. [CrossRef] [PubMed]

45. Dos Santos Marsico, V.; Lehmann, R.B.; de Assis Claro, C.A.; Amaral, M.; Vitti, R.P.; Neves, A.C.C.; da Silva Concilio, L.R. Three-dimensional finite element analysis of occlusal splint and implant connection on stress distribution in implant–supported fixed dental prosthesis and peri-implantal bone. *Mater. Sci. Eng. C* **2017**, *80*, 141–148. [CrossRef] [PubMed]

46. Gümrükçü, Z.; Korkmaz, Y.T.; Korkmaz, F.M. Biomechanical evaluation of implant-supported prosthesis with various tilting implant angles and bone types in atrophic maxilla: A finite element study. *Comput. Biol. Med.* **2017**, *86*, 47–54. [CrossRef] [PubMed]

47. Sevimay, M.; Turhan, F.; Kiliçarslan, M.A.; Eskitascioglu, G. Three-dimensional finite element analysis of the effect of different bone quality on stress distribution in an implant-supported crown. *J. Prosthet. Dent.* **2005**, *93*, 227–234. [CrossRef] [PubMed]

48. Haug, R.; Fattahi, T.; Goltz, M. A biomechanical evaluation of mandibular angle fracture plating techniques. *J. Oral Maxillofac. Surg.* **2001**, *59*, 1199–1210. [CrossRef] [PubMed]

49. Maló, P.; Nobre Mde, A.; Lopes, A.; Ferro, A.; Moss, S. Five-year outcome of a retrospective cohort study on the rehabilitation of completely edentulous atrophic maxillae with immediately loaded zygomatic implants placed extra-maxillary. *Eur. J. Oral Implantol.* **2014**, *7*, 267–281. [PubMed]

50. Migliorança, R.; Coppedê, A.; Dias Rezende, R.; de Mayo, T. Restoration of the edentulous maxilla using extrasinus zygomatic implants combined with anterior conventional implants: A retrospective study. *Int. J. Oral Maxillofac. Implants* **2011**, *26*, 665–672. [PubMed]

Article

Corrosion Study of Implanted TiN Electrodes Using Excessive Electrical Stimulation in Minipigs

Suzan Meijs [1,*], Kristian Rechendorff [2], Søren Sørensen [2] and Nico J.M. Rijkhoff [1]

[1] Department of Health, Science and Technology, Center for Sensory-Motor Interaction (SMI), Aalborg University, 9220 Aalborg, Denmark; nr@hst.aau.dk

[2] Materials Department, Danish Technological Institute, 8000 Århus, Denmark; krr@teknologisk.dk (K.R.); soren.steenfeldt.moller-sorensen@LEGO.com (S.S.)

* Correspondence: smeijs@hst.aau.dk

Received: 27 February 2019; Accepted: 25 March 2019; Published: 28 March 2019

Abstract: (1) Background: Titanium nitride (TiN) electrodes have been used for implantable stimulation and sensing electrodes for decades. Nevertheless, there still is a discrepancy between the in vitro and in vivo determined safe charge injection limits. This study investigated the consequences of pulsing implanted electrodes beyond the in vivo safe charge injection limits. (2) Methods: The electrodes were implanted for a month and then pulsed at 20 mA and 50 mA and 200 Hz and 400 Hz. Afterwards, the electrodes were investigated using electrochemical and analytical methods to evaluate whether electrode degradation had occurred. (3) Results: Electrochemical tests showed that electrodes that pulsed at 20 mA and 200 Hz (lowest electrical dose) had a significantly lower charge injection capacity and higher impedance than the other used and unused electrodes. (4) Conclusions: The electrodes pulsed at the lowest electrical dose, for which no tissue damage was found, appeared to have degraded. Electrodes pulsed at higher electrical doses for which tissue damage did occur, on the other hand, show no significant degradation in electrochemical tests compared to unused implanted and not implanted electrodes. It is thus clear that the tissue surrounding the electrode has an influence on the charge injection properties of the electrodes and vice versa.

Keywords: implanted electrodes; electrical stimulation; corrosion

1. Introduction

Titanium nitride (TiN) has been used for implantable electrodes for many decades, starting with cardiac pacing electrodes [1]. The demands on cardiac pacing electrodes increased when it was desired to sense the heart rhythm, in order to provide rate-adaptive pacing [2]. Despite the high voltages applied during cardiac pacing, the electrode polarization should remain low so that the heart signal can reliably be recorded [1,3]. Porous electrodes were highly desirable for that purpose [4] but the electrodes should also be biocompatible [1] and corrosion resistant [3].

At the end of the previous century, TiN also received interest as a material for neural stimulation and recording electrodes [5]. Neural stimulation and recording applications within this field include, among others, visual prosthesis [6], brain implants [7,8] and cochlear implants [9]. Initially, studies reported conflicting results [5,6,9], which was likely due to differences in the fabrication method [4]. The majority of studies, however, reported very favourable properties of porous TiN [5,6], which were due to its large surface area rather than specific material properties [4].

The performance of stimulation and recording electrodes can be evaluated using their safe charge injection limits (Q_{inj}), charge storage capacity (CSC) and impedance. Q_{inj} is evaluated by comparing the electrode polarization under pulsing conditions to the safe potential limits established using slow sweep cyclic voltammetry (CV). The safe potential window is typically defined by the potentials at which water reduction and oxidation occurs. CSC is a measure of how much charge can be stored on the surface of the electrode and is measured using CV. The amount of charge available during fast pulsing, however, is typically much less than CSC. Impedance magnitude (typically at 1 kHz) can be used as a measure for battery consumption or recording performance. The lower the impedance, the better [10].

These properties are typically investigated under in vitro conditions in inorganic saline [5,6,9–20]. However, CSC, Q_{inj} and the impedance spectrum differ under acute and chronic in vivo circumstances [4,7,8,21–28]. Q_{inj} and electrode polarization have been reported to be significantly lower after implantation compared to in vitro measurements [7,8,23–27]. Moreover, they have been reported to decrease during the implanted period, when electrode failure does not occur [7,23–26].

TiN has long been known as a biocompatible [29–31] and corrosion resistant material [3,32,33], even under cathodic high voltage pulsing conditions [34]. Under anodic conditions, TiN oxidation reactions may occur, which primarily lead to passivation of these reactions until higher anodic voltages are reached [32]. At very high anodic voltages, TiN will eventually be degraded [34]. However, as Q_{inj} is lower when implanted compared to in vitro [7,8,21–27], unsafe voltages may be reached during pulsing. The aim of this study was therefore to investigate whether implanted TiN electrodes would degrade during pulsing when Q_{inj} measured in vivo was exceeded but Q_{inj} measured in saline was not.

2. Materials and Methods

Four Göttingen minipigs were implanted with four working electrodes (electrode pins) and four large surface area pseudo-reference disk electrodes. Minipigs were selected because the subcutaneous adipose tissue is similar to adipose tissue in humans. The number of electrodes and pseudo-reference electrodes was chosen in order not to cause excessive discomfort to the animals and thereby also to increase the homogeneity in the results. The electrodes were made from Ti6Al4V and coated with porous TiN. The animals recovered from anaesthesia and were monitored for one month before the corrosion experiments were conducted. The work was carried out according to Danish and European legislation (ethical approval license no: 2014-15-0201-00268).

2.1. Electrode Fabrication

TiN coatings were deposited on electrode pins (6 mm^2) made of a Ti6Al4V alloy and Ti disks (1000 mm^2) by reactive magnetron sputtering on a CC800/9 SiNOx coating unit (CemeCon AG, Würselen, Germany). The coatings were sputtered from four Ti targets (88 $\times$ 500 mm^2) with 99.5% purity in a Ar/N$_2$ mixture atmosphere. The purity of the gases was 99.999% and the Ar/N$_2$-flow was 300 sccm/350 sccm. The deposition time was 21,000 s. The electrodes underwent three-fold rotation during the coating process.

The electrodes coatings were investigated after deposition using samples taken from the same batch. Scanning electron microscopy (SEM) (Nova 600, FEI Company, Hillsboro, OR, USA) images were taken at magnifications ranging from 450$\times$ to 25,000$\times$ to get an overview of the surface and to investigate the surface structure of the electrodes in detail. A silicium sample coated in the same process was used to study the thickness, homogeneity and porous structure of the TiN coating using SEM. Images were recorded at a magnification of 40,000$\times$.

An ethylene tetrafluoroethylene (ETFE) coated 35N LT wire (Heraeus, Yverdon, Switzerland) was crimped to the hollow end of the electrode pins. A polyethether ketone (PEEK) body and silicone tines were produced using injection moulding to insulate the electrode pins. The tines were first glued to the PEEK body using a silicone adhesive. The PEEK body with tines was then glued to the electrode pins also using a silicone adhesive. Further details and figures of the electrode production can be found in Reference [24]. The electrodes were cleaned thoroughly before they were sterilized by an overdose of electron-beam processing.

2.2. Surgical Procedure

For electrode implantation, the animals were anesthetized using Propofol. Small incisions were made in the back and the electrodes were implanted using a custom-made implantation tool. This ensured that the electrodes were placed in tight pockets in subcutaneous adipose tissue, which promotes fast healing and ingrowth. Five electrodes were implanted into each pig, four of which were used for intense electrical stimulation and one electrode in each pig served as a control. Four counter electrode disks with a percutaneous wire were also implanted in each pig. The minipigs recovered from the procedure and were carefully monitored in order to detect and treat cases of infection. The electrodes were not used for one month until the pigs were anaesthetized again using sevoflurane to perform electrical stimulation. After the stimulation sessions were completed, the electrodes were carefully dissected from the tissue. The electrodes were extensively cleaned using demineralized water and alcohol, they were then rinsed and stored dry, so that they could be further investigated.

2.3. Electrical Stimulation

Electrical stimulation was performed using a DS5 (Digitimer, Hertfordshire, UK) per electrode. The device was shorted between the pulses using a custom-build set-up to prevent drifting of the baseline potential. Biphasic, charge balanced 200 µs square pulses were applied, cathodic first with an inter-phase interval of 40 µs during which no current was applied. Stimulation was performed for 6 h in total, divided into three 2-h sessions. Before, between and after these sessions, voltage transient measurements (VTM) were recorded for each electrode using a VersaSTAT 3 potentio-galvanostat (Princeton Applied Research, Oak Ridge, TN, USA).

Four stimulation paradigms were applied:

- Group 1: 20 mA, 200 Hz
- Group 2: 20 mA, 400 Hz
- Group 3: 50 mA, 200 Hz
- Group 4: 50 mA, 400 Hz

During pilot experiments, it was verified that the group 1 stimulation paradigm did not cause tissue damage after one week of implantation (see Figure S1). To cause electrode damage, we decided to increase 2 parameters: stimulation frequency and stimulation amplitude. The stimulation frequency was doubled, which was expected to cause electrode damage due to an increasing trend in the inter-pulse potential [35]. The stimulation amplitude was set to the maximum the DS5 can deliver, which was expected to cause electrode damage by increasing the electrode potential during stimulation. It was expected that group 2–4 protocols would result in tissue damage; therefore photographs were taken of the tissue surrounding the electrodes to document the amount of tissue damage. However, the focus of this study is corrosion and the electrodes were thus investigated more extensively using electrochemical and analytical methods.

During stimulation, the voltage transients were recorded every 30 min using an oscilloscope. From these voltage transients, the resistive drop after pulse cessation (IR-drop) was calculated as [10]:

$$\text{IR-drop} = E_{\text{pulse\_end}} - E_{\text{pulse\_end}+40} \tag{1}$$

where $E_{\text{pulse\_end}}$ is the recorded potential at the end of the cathodic pulse and $E_{\text{pulse\_end}+40}$ is the potential 40 µs after pulse cessation (see Figure 1) [10]. E_{mc} and E_{ma} are the maximum cathodic and anodic voltage excursions after IR-drop is subtracted from the original voltage transient.

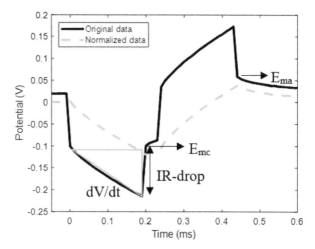

Figure 1. IR-drop, dV/dt, E_{mc} and E_{ma} are derived from the original data, while in the manuscript normalized data are presented. The data is normalized by subtracting the IR-drop and setting the pre-pulse potential to 0.

VTM before, between and after the stimulation blocks, was performed using the VersaSTAT 3 potentio-galvanostat using the same stimulation pulse.

The pulsing capacitance (C_{pulse}) was computed using the slope (dV/dt) of the voltage transient:

$$I_{stim} = C_{pulse} \cdot \frac{dV}{dt} \tag{2}$$

where I_{stim} is the stimulation current (1 mA while implanted and 5 mA in the electrochemical characterization). Q_{inj} was calculated using the current (I_{max}) at which E_{mc} or E_{ma} reached the safe potential limits (-0.6 and 0.9 V vs. open circuit potential, respectively) [10,24]:

$$Q_{inj} = \frac{I_{max} \cdot t}{A} \tag{3}$$

where t is the pulse duration (200 µs) and A is the geometrical surface area of the electrodes (6 mm^2). When voltage excursions exceeded machine limits (± 10 V), I_{max} was extrapolated from the highest current assuming a linear relation.

$$V_{ext} = V_m \left(1 + \frac{I_{ext} - I_m}{I_m}\right) \tag{4}$$

where V_m and I_m were the measured potential and current, respectively, and V_{ext} and I_{ext} were the extrapolated potential and current. When V_{ex} reached the potential limits, I_{ext} was used as I_{max} in (3). This method provided accurate results using data for which I_{max} was measured.

2.4. Coating Characterization

After explantation, all electrodes were characterized using SEM and energy-dispersive x-ray spectroscopy (EDX). The electrochemical properties were investigated using electrochemical impedance spectroscopy (EIS), CV and VTM. Two electrode groups were added to the 4 groups of active implants, therefore these measurements were performed on six electrode groups:

- Group 1: implanted—20 mA, 200 Hz
- Group 2: implanted—20 mA, 400 Hz

- Group 3: implanted—50 mA, 200 Hz
- Group 4: implanted—50 mA, 400 Hz
- Group 5: implanted controls
- Group 6: not implanted controls

SEM (Nova 600, FEI Company, Hillsboro, OR, USA) images were recorded at magnifications varying from $450\times$ to $10000\times$ to obtain an overview of the surface and to investigate in detail the surface structure of the electrodes. SEM images of the not implanted control electrodes (group 6) were made, both to compare to the other electrode groups, as well as to investigate the uniformity of the coating after deposition. Further SEM analysis was carried out on a Si-wafer which was coated in the same process as the electrodes. The Si-wafer was mounted in a manner similar to the electrodes and the measured thickness is representative for the coating thickness on the electrodes. The advantage of using Si-wafer is that a cross-section analysis of the coating can be done easily. EDX (EDAX, AMETEK, Leicester, UK) spectra were made to investigate the chemical composition of the coatings after deposition and to determine whether the surface chemistry of the electrodes changed after having been implanted and after intense pulsing.

Electrochemical characterization measurements were performed in an electrochemical cell at room temperature using phosphate buffered saline as the electrolyte. The measurements were performed in a 3-electrode set-up, using the above mentioned porous TiN electrodes as working electrodes (0.06 cm^2), a Ag⏐AgCl reference electrode (1.6 cm^2) and a platinum foil counter electrode (50 cm^2).

Solartron, Model 1294 in conjunction with 1260 Impedance/gain-phase Analyzer (Solartron Analytical, Farnborough, UK) were used to perform EIS measurements. Accompanying SMaRT software was used to run the measurements. A sinusoidal current was used at frequencies from 0.1 Hz to 100 kHz, with 10 measurements per decade. Three different currents (5, 10 and 50 µA) were used to ensure that the measurement currents were in the linear operation range of the electrode [36]. An integration time of 10 s was used to obtain a reliable and noise-free signal.

Cyclic voltammetry (CV) was performed by cycling the electrode potential was cycled between the safe potential limits (-0.6 and 0.9 V vs. Ag⏐AgCl) previously established for similar electrodes [24]. The sweep rates used for CV were 0.05, 0.1, 0.5 and 1.0 V/s. Ten cycles were made at each sweep rate, the last cycle was used for data analysis. The cathodic charge storage capacity (CSC) was derived from the CV by taking the integral of the CV below the zero-current axis [10].

VTM were conducted in the same manner as described above for the implanted electrodes, except the 3-electrode setup and the electrochemical cell were employed. The maximum charge injection limit (Q_{inj}) and pulsing capacitance (C_{pulse}) were derived according to Equations (1)–(3).

2.5. Statistics

The data recorded during the 2-h pulsing sessions using an oscilloscope were filtered using a low-pass Butterworth filter (passband 5 kHz, stopband 15 kHz). E_{mc} and IR-drop were then normalized to the first measurement (session 1, start). E_{mc} was selected for statistical analysis to represent the electrode polarization and IR_{drop}, as a measure of the tissue resistance. Before, between and after the pulsing sessions, voltage transients were recorded using the VersaSTAT 3 (Princeton Applied Research, Oak Ridge, TN, USA). From these voltage transients Q_{inj} and C_{pulse} were used to further quantify the electrochemical performance of the electrodes. A linear mixed model was used to statistically analyse the data. Parameters (group, session, time and combinations thereof) were added stepwise to the model, until adding another parameter did not make a significant difference to the model.

One-way ANOVA was used to investigate the electrochemical properties of the electrodes after explantation (electrochemical cell setup). The following electrochemical properties of the 6 different electrode groups were used for statistical analysis:

- Cathodic CSC at 0.05 and 1.0 V/s
- Impedance magnitude at 0.1 Hz

- Q_{inj}

Significant findings are reported at p-values smaller than 0.05.

3. Results

All implantations were carried out without any complications. The animals recovered well from the surgery and no infections were observed during the month the electrodes were implanted.

3.1. General Coating Characteristics

In contrast to the well-known yellow-golden coloured TiN, the coatings on the electrodes had a brownish colour. To analyse the structure and chemical composition of the coating, the electrodes were studied in SEM and EDX. The overview SEM image (Figure 2a) shows a uniform coating on the electrode and in the corresponding EDX spectrum (Figure 2b) the expected peaks belonging to Ti and N are present. The quantification of the amounts of Ti and N from an EDX spectrum is difficult because the K-line of N and the L-line of Ti are very close. Here, numbers close to a 1:1 atomic ratio of Ti to N are found (note that weight-% is used in Figure 2b). The SEM images in Figure 3a clearly show a faceted structure, typical for TiN deposited at high pressure. The cross-section SEM image in Figure 3b shows the porous morphology of the coating as well. The coating thickness is approximately 6 µm.

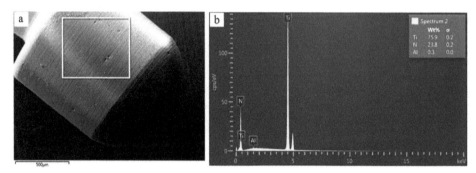

Figure 2. (**a**) Scanning electron microscope (SEM) image of the coated electrode. The indicated region is the area over which EDX was performed. The scale-bar is 500 µm. (**b**) Typical energy dispersive X-ray spectroscopy (EDX) spectrum corresponding to the area indicated in (**a**).

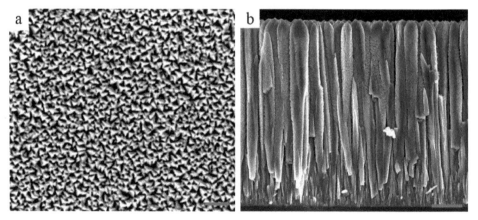

Figure 3. (**a**) Top-view (magnification: 25,000×) and (**b**) cross-section SEM images (magnification: 40,000×) of TiN coating on Si-wafer. Scale bar: 1 µm.

3.2. Changes in Electrochemical Properties during Intensive Pulsing

The shorting part of the setup broke down during the last series of measurements. The last two stimulation sessions could therefore not be completed with one of the electrodes in the 20 mA—00 Hz group. The data obtained with this electrode after the breakdown was not used in the analyses.

The significant parameters of the statistical model for IR-drop were: Time, Session, Group × Session and Time × Session. For E_{mc}, Group was an additional significant parameter of the statistical model. Figure 4 shows that the results for IR-drop and E_{mc} were similar. Both IR-drop and E_{mc} were significantly larger during session 1 compared to sessions 2 and 3 for electrode groups 2, 3 and 4. IR-drop and E_{mc} were only significantly larger during session 1 compared to sessions 2 and 3 at the 30 and 60 min measurements. Figure 4a,b also show that IR-drop (for groups 2, 3 and 4) and E_{mc} (all groups) were significantly larger after 30 and 60 min of pulsing compared to after 90 and 120 min of pulsing during session 1.

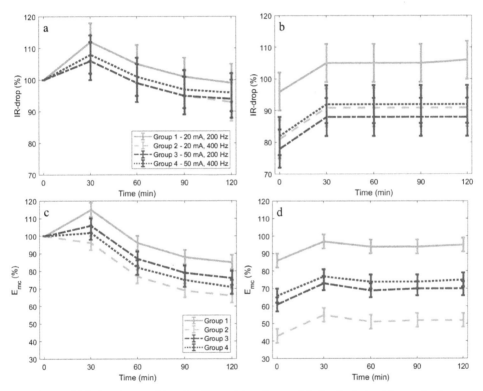

Figure 4. (**a**) IR-drop increased from baseline and then decreased during the first pulsing session for electrode groups 2, 3 and 4. (**b**) During session 2 and 3 (shown), an increase in IR-drop was seen from the start of stimulation, after which IR-drop remained stable. The IR-drop of group 2, 3 and 4 electrodes was significantly smaller during sessions 2 and 3 compared to session 1. (**c**) The same trend was observed for E_{mc} of all electrode groups but to an even greater extent (notice the axis). (**d**) An increase in E_{mc} was also observed from the start of stimulation during sessions 2 and 3 (shown). E_{mc} of group 2, 3 and 4 electrodes was also significantly smaller during sessions 2 and 3 compared to session 1 but notice again the difference in the axis of IR-drop and E_{mc}.

E_{mc} of group 1 electrodes was significantly larger than E_{mc} of group 2, 3 and 4 electrodes in all sessions. During session 2, E_{mc} of group 2 electrodes was significantly smaller than E_{mc} of group 1 and 3 electrodes and during session 3 E_{mc} of group 2 electrodes was significantly smaller than E_{mc} of group

1, 3 and 4 electrodes. Figure 4c,d show that IR-drop and E_{mc} of all electrode groups was significantly smaller at the start of stimulation compared to after 30, 60, 90 and 120 min of pulsing during session 3. The same was found for session 2.

3.3. Changes in Electrochemical Properties between Pulsing Sessions

For both C_{pulse} and Q_{inj}, the significant fixed effects were: Time, Group and Time*Group. The results of the statistical analysis for C_{pulse} and Q_{inj} were identical, except for a baseline difference between electrode groups observed for Q_{inj} (p = 0.045). Q_{inj} of group 2 electrodes ($8.3 \pm 2.4\ \mu C/cm^2$) was significantly smaller than Q_{inj} of group 3 and 4 electrodes ($15.8 \pm 2.4\ \mu C/cm^2$).

Figure 5a,b show that Q_{inj} and C_{pulse}, respectively, of group 1 electrodes did not change significantly. For group 2, 3 and 4 electrodes, Q_{inj} increased significantly to values of $26.45 \pm 2.7\ \mu C/cm^2$, $50.00 \pm 2.4\ \mu C/cm^2$ and $52.50 \pm 2.4\ \mu C/cm^2$, respectively, after 6 h of intense pulsing. C_{pulse} of electrode groups 2, 3 and 4 increased significantly to capacitances of $54.1 \pm 5.4\ \mu F/cm^2$, $97.1 \pm 4.9\ \mu F/cm^2$ and $106.0 \pm 4.9\ \mu F/cm^2$, respectively. Figure 5c,d show that the increase in C_{pulse} caused a decrease in electrode polarization. This decrease in electrode polarization led to an increased Q_{inj}.

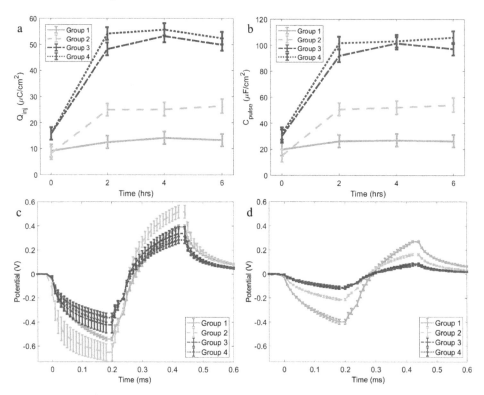

Figure 5. (a) After the first pulsing session Q_{inj} was increased compared to before pulsing for electrode groups 2, 3 and 4. (b) The same was observed for C_{pulse} of electrode groups 2, 3 and 4. (c) Normalized voltage transients recorded before pulsing at 3 mA. (d) Normalized voltage transients recorded at 3 mA after 2 h of pulsing.

Electrode group 1 had a significantly smaller Q_{inj} than all other electrode groups after the first pulsing session, which remained after the second and third pulsing session. Furthermore, electrode group 2 had a significantly smaller Q_{inj} than electrode groups 3 and 4 after the first pulsing session.

This difference also remained significant after pulsing sessions 2 and 3. The same group differences were observed for C_{pulse}.

3.4. Electrochemical Characteristics after Explantation

The results of the electrochemical characterization in phosphate buffered saline were largely consistent across measurements, as shown in Figure 6. Group 1 electrodes had a significantly smaller CSC at 0.05 and 1.0 V/s and a significantly larger impedance magnitude at 0.1 Hz compared to all other electrode groups. They also had a significantly smaller Q_{inj} compared to group 2 electrodes and the control electrodes in groups 5 and 6. But the Q_{inj} of group 1 electrodes was not significantly different from group 3 and 4 electrodes.

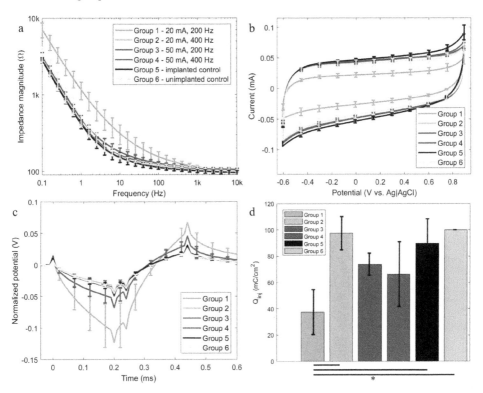

Figure 6. (**a**) The impedance magnitude was significantly larger for electrode group 1 compared to the other electrode groups, which was implanted and pulsed at the lowest electrical dose. (**b**) The cyclic voltammogram shows that the charge storage capacity of electrode group 1 was significantly smaller than the other electrode groups. (**c**) The normalized voltage transients at 5 mA show that the slope of the group 1 electrodes was larger than the slopes of the other electrode groups but no significant difference was found for C_{pulse}. (**d**) Q_{inj} of group 1 electrodes was significantly smaller than Q_{inj} of group 2, 5 and 6 electrodes.

3.5. Coating Properties after Explantation

SEM images (see Figure 7) showed that the electrode surfaces were intact after 6 h of intense stimulation. The coatings were all undamaged and had the same faceted structure, which is typical for TiN, as the electrodes had after deposition of the coating. Figure 7 indicates that there were no differences in chemical composition of group 1 and group 4 electrodes. The EDX spectra of all electrodes in all groups showed similar levels of titanium, nitrogen and oxide.

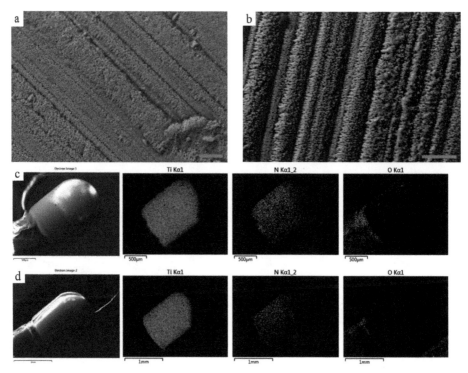

Figure 7. (**a**) Close-up SEM image (magnification: 10,000×) of an electrode in group 1 (20 mA—200 Hz), showing that the surface structure is still intact (scale bar: 10 µm). The same is true for (**b**), showing the detailed structure of an electrode in group 4 (50 mA—400 Hz, scale bar: 10 µm, magnification: 15,000×). (**c**) Overview SEM (magnification 450×) and corresponding EDX images of an electrode in group 1 show that the chemical composition of the electrode surface after pulsing was very similar to (**d**) the chemical composition of the electrode surface of an electrode in group 4. Also visualized by an overview SEM (magnification 450×) and corresponding EDX images of an electrode in group 4.

4. Discussion

Several previous studies [37–39] have shown that TiN coatings grown at high N-partial pressures become porous. These coatings have a brown colour that clearly distinguish them from the standard golden-coloured TiN used on tools and components. For use in electrode applications, it is essential to use the porous type of TiN coating as they have a high effective surface area, which in turn leads to a low impedance [3,4]. Cunha et al. [39] did a systematic study of the influence of N-content of the morphological structural and electrochemical properties of TiN coatings. In the case of high N-content (Ti:N ratio 1:1.34), these authors obtain results similar to ours regarding the coating morphology. The reason for the discrepancy in chemical composition (in our case we measure a Ti:N ratio 1:1 with EDX) could be that Cunha et al. [39] have determined the chemical composition using Rutherford Backscattering Spectrometry, which provides a more precise determination compared to EDX. Other studies have found that porous near-stoichiometric TiN coatings can be obtained just by adjusting the energy available during film growth [40].

Six groups of electrodes were used in this study, five of which were implanted and four of which were used for 6 h of intense pulsing at different electric doses:

- Group 1: 20 mA, 200 Hz
- Group 2: 20 mA, 400 Hz

- Group 3: 50 mA, 200 Hz
- Group 4: 50 mA, 400 Hz
- Group 5: implanted controls
- Group 6: un-implanted controls

During intense pulsing, IR-drop was derived from the voltage transients as a measure of tissue impedance and E_{mc} was used as a measure of electrode polarization. Between the three 2-h pulsing sessions, Q_{inj} was evaluated and from these voltage transients C_{pulse} was derived.

No significant changes were observed during intense pulsing for group 1 electrodes, receiving the lowest electrical dose. For group 2, 3 and 4 electrodes the IR-drop and E_{mc} decreased and Q_{inj} and C_{pulse} increased. The stability in pulsing properties of group 1 electrodes together with the fact that these electrodes received the lowest electrical dose, would intuitively lead to the expectation that these electrodes did not corrode [4,7,8,23–26]. For group 2, 3 and 4 electrodes, on the other hand, it could be expected that corrosion may have occurred, even though the observed electrochemical changes are favourable in the light of pulsing capability (higher charge injection, lower electrode polarization). Passivation at high anodic potentials was the main expected corrosive reaction [33,34], leading to decreased pulsing capability. Excessive bubbling due to water reduction [34], however, may lead to cracking of the coating. This could increase the surface area and thereby lead to an apparent increase Q_{inj}. The results of the electrochemical characterizations after explantation showed that group 1 electrodes had significantly deteriorated electrochemical properties compared to all other electrode groups. Group 2, 3 and 4 electrodes, on the other hand, had no different electrochemical properties after explantation than the two control groups (5 and 6).

Analytical investigations could neither confirm nor reject the electrochemical results. SEM images show that all coatings seemed to be intact. EDX spectra did not reveal differences between the harshest and mildest stimulated electrodes either. However, it must be noted that it is difficult to distinguish between the oxygen and nitrogen signal using EDX because the K-peaks of the two elements are quite close. As oxidation of the coating could have occurred, other analytical methods have been attempted (X-ray photoelectron spectroscopy and Time-of-Flight Secondary Ion Mass Spectrometry) in order to detect any differences in oxygen amounts. Preliminary results were unsuccessful, mainly because the geometry and size of the electrode is quite challenging in both the experimental set-ups. It is, however, very unlikely that the coating oxidized without showing any signs of damage. Norlin et al. [34] show SEM images porous TiN electrodes after anodic pulsing, which are severely damaged. For the current electrodes no signs of damage were found using SEM and EDX analysis.

Corrosion studies of stimulation electrodes have been performed extensively in saline [3,34,41–43]. Some studies found that the damage threshold is exactly at the water window limit [43], others suggest that the water window limit may be too conservative under pulsing conditions [41,44] and yet others suggest that corrosion may occur even within the limits of water reduction and oxidation [42]. TiN has been most extensively investigated by Norlin et al. [3,34,45]. Their pulsing study was carried out using 700V pulses of both anodic and cathodic polarity [34]. As expected, TiN showed severe corrosion upon anodic pulsing but was stable when cathodic pulses were applied. The voltages recorded in this study during constant current pulsing were, however, a factor 15–30 smaller. In a later study, the electrodes were aged using more conservative voltages (−3 and 1 V vs. Ag | AgCl), corresponding to 4 months of use based on the charge passed [3]. TiN proved very stable, which was expected, as high anodic voltages were avoided. In the current study, very high anodic voltages were also avoided by using a cathodic first stimulation paradigm. Anodic voltages between 1 and 3 V (vs. open circuit potential) were observed with the highest voltages in the 50 mA groups (groups 3 and 4). No signs of corrosion were observed for the electrodes in those groups, while corrosion of the anodically pulsed samples by Norlin et al. [34] was obvious in SEM images.

It has also been shown before that safe limits obtained in inorganic solutions do not necessarily apply to electrodes in protein containing solution [46] and implanted electrodes [47,48]. It was therefore concluded that proteins must protect the electrode surface against corrosion [46,48,49]. However,

Q_{inj} was never measured in vivo for these electrodes and it is therefore unclear whether or not it was exceeded [47,48]. The in vitro water window limits for platinum were not exceeded in either of the studies but in both studies corrosion was observed nevertheless [47,48]. Shepherd et al. [48] argue, however, that corrosion was not stimulation-induced but due to production failures. In the current study, Q_{inj} (measured in vivo) was exceeded for all stimulated electrode groups (1, 2, 3 and 4) during all stimulation sessions. However, group 1 and 2 electrodes were pulsed at approximately 20% of Q_{inj} as measured in saline and groups 3 and 4 electrodes were pulsed at approximately 50% of the in vitro Q_{inj}. Robblee et al. [47] stimulated their electrodes at approximately 5% and 30% of Q_{inj} in vitro. They observed platinum dissolution for all electrodes pulsed at 30% of Q_{inj} (in vitro) and less for electrodes pulsed at 5%. Shepherd et al. [48] stimulated the electrodes at 5-10% of Q_{inj} in vitro but concluded that the observed corrosion was not stimulation-induced. This makes it obvious that in vitro safe limits cannot be applied in vivo. However, it does not rule out that limits measured in vivo using techniques developed in vitro may be too conservative.

Interestingly, we found that corrosion most likely occurred for the electrodes pulsed at the lowest electrical dose (group 1 electrodes; 20% of Q_{inj} in vitro and 200 Hz). As group 2, 3 and 4 electrodes showed no signs of corrosion in the electrochemical characterizations after explantation, the occurrence of corrosion seems not only dependent on the electrode potentials or charge delivered. We suspect that the occurrence of corrosion is not only medium dependent (organic vs. inorganic, basic vs. acidic) but also tissue dependent. Although tissue damage was not the focus of this study, it seems to play an important role. No tissue damage seems to have occurred for electrode group 1, while tissue damage with increasing severity occurred for electrode groups 2–4 (see Figure S2). The lack of tissue damage for electrode group 1 is confirmed by the lack of change in IR-drop, which is representative of tissue impedance [10]. The same amount of charge density per phase was injected for electrode groups 1 and 2 but due to the increased frequency tissue damage is likely to have occurred in group 2 [50,51]. Tissue damage obviously occurred in electrode groups 3 and 4 (see Figure S2). Based on the IR-drop data, it appears that the tissue was damaged during the first hour of the first pulsing session for electrode groups 2, 3 and 4. It seems that a new electrode-tissue interface was formed that remained during pulsing session 2 and 3. This new electrode-tissue interface allowed for more charge injection, as Q_{inj} and C_{pulse} were significantly increased after the first pulsing session compared to before pulsing. And although the electrodes were still pulsed beyond the increased Q_{inj} in vivo, the formation of a new electrode-tissue interface and corresponding increase in Q_{inj} may thus have prevented corrosion.

The electrode-tissue interface appears to play a very important role with regards to the occurrence of corrosion. These results can therefore only be applied to stimulation electrodes implanted in adipose tissue, like ours [23,24,52] and like Bion [53] for example. They cannot be applied to implants in the brain [47], the cochlea [48] or the blood stream [1,2,4]. Furthermore, our electrode is a macro-electrode (0.06 cm^2). There are indications that different charge injection limits apply to smaller microelectrodes [51]. These results might therefore not apply to microelectrodes. Lastly, as it is challenging to work with larger animals, such as minipigs, the number of animals is low compared to rodent studies for example. The results, however, are consistent across measurements and rather homogeneous within the electrode groups and were thus statistically significant.

With the recent increase in investment in "electroceuticals," the development of novel, smaller and more sophisticated implants may be anticipated [54]. It is therefore more important than ever before to establish safe limits that apply to these specific implants [51]. We show that this is not only relevant in the light of tissue damage but also with respect to corrosion and long-term electrochemical performance of the implants. In the light of tissue damage due to corrosion, TiN appears to be a very suitable material for implants. There seems to occur no dissolution of the material [55], like with Pt [46] and IrOx [43]. As long as very high anodic potentials are avoided [34], we show that no corrosion occurs even after almost 9 million pulses. When corrosion does occur, its product (a passivation layer) remains attached to the electrodes and is not harmful to the tissue [33,55].

5. Conclusions

It was long suspected that in vitro safe limits established for implantable electrodes may not be applicable in vivo, which we confirm here. We also show that the type of tissue in which the electrode is implanted has an influence on safety limits. Biocompatibility and corrosion resistance cannot be viewed as two separate properties of implantable stimulation (and sensing) electrodes. Tissue responses influence the electrochemical behaviour of implanted electrodes and use of the electrodes influences the tissue surrounding the electrodes. It is therefore of great importance that safe limits are established for each electrode depending on the tissue in which it will be implanted.

Supplementary Materials: The following are available online at http://www.mdpi.com/2075-4701/9/4/389/s1, Figure S1: (a) Tissue around the tip of an electrode stimulated at 20 mA-200 Hz after 1 week of implantation. The tissue was stained using haematoxylin and eosin (H&E), which is the most commonly applied stain in medical diagnostics. Some inflammatory cells can still be observed but capsule formation has begun to take place. (b) Cells around the silicone part of the electrode appear very similar to those around the electrode tip, indicating no signs of stimulation-induced tissue damage. Figure S2: (a) The tissue around electrodes in group 1 showed no signs of tissue damage upon sacrifice. (b) The tissue around electrodes in group 2 showed some redness around the electrode tip, which likely is due to tissue damage. (c) The tissue around the electrode tips of electrodes in group 3 showed obvious tissue damage but the tissue around the insulated parts was unaffected. (d) The tissue around the electrode tips of electrodes in group 4 showed even more extensive tissue damage and bleeding. Nevertheless, the tissue around the insulated parts was unaffected.

Author Contributions: Conceptualization, S.M. and N.R.; methodology, S.M., K.R., S.S. and N.R.; writing—original draft preparation, S.M.; writing—review and editing, S.M., K.R., S.S and N.R.; project administration, N.R.; funding acquisition, N.R.

Funding: This research was funded by the Danish National Advanced Technology Foundation.

Acknowledgments: The authors thank Neurodan A/S, a member of the Ottobock group, for supplying the electrodes used in this study.

Conflicts of Interest: The authors declare no conflict of interest.

References

1. Saldach, M.; Hubmann, M.; Weikl, A.; Hardt, R. Sputter-deposited TiN electrode coatings for superior pacing and sensing performance. *Pacing Clin. Electrophysiol.* **1990**, *13*, 1891–1895. [CrossRef]

2. Lau, C.P.; Tse, H.F.; Camm, A.J.; Barold, S.S. Evolution of pacing for bradycardias: Sensors. *Eur. Heart J. Suppl.* **2007**, *9*, I11–I22. [CrossRef]

3. Norlin, A.; Pan, J.; Leygraf, C. Investigation of electrochemical behavior of stimulation/sensing materials for pacemaker electrode applications: I Pt, Ti and TiN coated electrodes. *J. Electrochem. Soc.* **2005**, *152*, J7–J15. [CrossRef]

4. Hubmann, M.; Bolz, A.; Hartz, R.; Saldach, M. Long term performance of stimulation and sensing behaviour of TiN and Ir coated pacemaker lead having a fractal surface structure. In Proceedings of the 1992 14th Annual International Conference of the IEEE Engineering in Medicine and Biology Society, Paris, France, 29 October–1 November 1992; Volume 6.

5. Janders, M.; Egert, U.; Stelzle, M.; Nisch, W. Novel thin film titanium nitride micro-electrodes with excellent charge transfer capability for cell stimulation and sensing applications. In Proceedings of the 1996 18th Annual International Conference of the IEEE Engineering in Medicine and Biology Society, Amsterdam, The Netherlands, 31 October–3 November 1 1996; Volume 1.

6. Zhou, D.M.; Greenberg, R.J. Electrochemical characterization of titanium nitride microelectrode arrays for charge-injection applications. In Proceedings of the 2003 25th Annual International Conference of the IEEE Engineering in Medicine and Biology Society, Cancun, Mexico, 17–21 September 2003; Volume 2.

7. Kane, S.R.; Cogan, S.F.; Ehrlich, J.; Plante, T.D.; McCreery, D.B.; Troyk, P.R. Electrical performance of penetrating microelectrodes chronically implanted in cat cortex. *IEEE Trans. Biomed. Eng.* **2013**, *60*, 2153–2160. [CrossRef] [PubMed]

8. Brunton, E.K.; Winther-Jensen, B.; Wang, C.; Yan, E.B.; Hagh Gooie, S.; Lowery, A.J.; Rajan, R. In vivo comparison of the charge densities required to evoke motor responses using novel annular penetrating microelectrodes. *Front. Neurosci.* **2015**, *9*, 265. [CrossRef]

9. Weiland, J.D.; Anderson, D.J.; Humayun, M.S. In vitro electrical properties for iridium oxide versus titanium nitride stimulating electrodes. *IEEE Trans. Biomed. Eng.* **2002**, *49*, 1574–1579. [CrossRef]

10. Cogan, S.F. Neural stimulation and recording electrodes. *Annu. Rev. Biomed. Eng.* **2008**, *10*, 275–309. [CrossRef] [PubMed]

11. Cogan, S.F.; Troyk, P.R.; Ehrlich, J.; Plante, T.D. In vitro comparison of the charge-injection limits of activated iridium oxide (AIROF) and platinum-iridium microelectrodes. *IEEE Trans. Biomed. Eng.* **2005**, *52*, 1612–1614. [CrossRef] [PubMed]

12. Cogan, S.F.; Troyk, P.R.; Ehrlich, J.; Plante, T.D.; Detlefsen, D.E. Potential-biased, asymmetric waveforms for charge-injection with activated iridium oxide (AIROF) neural stimulation electrodes. *IEEE Trans. Biomed. Eng.* **2006**, *53*, 327–332. [CrossRef] [PubMed]

13. Whalen, J.J.; Young, J.; Weiland, J.D.; Searson, P.C. Electrochemical characterization of charge injection at electrodeposited platinum electrodes in phosphate buffered saline. *J. Electrochem. Soc.* **2006**, *153*, C834–C839. [CrossRef]

14. Cogan, S.F.; Troyk, P.R.; Ehrlich, J.; Gasbarro, C.M.; Plante, T.D. The influence of electrolyte composition on the in vitro charge-injection limits of activated iridium oxide (AIROF) stimulation electrodes. *J. Neural Eng.* **2007**, *4*, 79–86. [CrossRef] [PubMed]

15. Cogan, S.F.; Ehrlich, J.; Plante, T.D.; Smirnov, A.; Shire, D.B.; Gingerich, M.; Rizzo, J.F. Sputtered iridium oxide films for neural stimulation electrodes. *J. Biomed. Mater. Res. B* **2009**, *89*, 353–361. [CrossRef]

16. Cogan, S.F.; Ehrlich, J.; Plante, T.D. The effect of electrode geometry on electrochemical properties measured in saline. In Proceedings of the 2014 36th Annual International Conference of the IEEE Engineering in Medicine and Biology Society (EMBC), Chicago, IL, USA, 26–30 August 2014.

17. Boehler, C.; Stieglitz, T.; Asplund, M. Nanostructured platinum grass enables superior impedance reduction for neural microelectrodes. *Biomaterials* **2015**, *67*, 346–353. [CrossRef] [PubMed]

18. Weremfo, A.; Carter, P.; Hibbert, D.B.; Zhao, C. Investigating the interfacial properties of electrochemically roughened platinum electrodes for neural stimulation. *Langmuir* **2015**, *31*, 2593–2599. [CrossRef]

19. Ghazavi, A.; Cogan, S.F. Electrochemical characterization of high frequency stimulation electrodes: Role of electrode material and stimulation parameters on electrode polarization. *J. Neural Eng.* **2018**, *15*, 036023. [CrossRef]

20. Deku, F.; Joshi-Imre, A.; Mertiri, A.; Gardner, T.J.; Cogan, S.F. Electrodeposited Iridium Oxide on Carbon Fiber Ultramicroelectrodes for Neural Recording and Stimulation. *J. Electrochem. Soc.* **2018**, *165*, D375–D380. [CrossRef]

21. Wei, X.F.; Grill, W.M. Impedance characteristics of deep brain stimulation electrodes in vitro and in vivo. *J. Neural Eng.* **2009**, *6*, 046008. [CrossRef] [PubMed]

22. Leung, R.T.; Shivdasani, M.N.; Nayagam, D.A.; Shepherd, R.K. In vivo and in vitro comparison of the charge injection capacity of platinum macroelectrodes. *IEEE Trans. Biomed. Eng.* **2015**, *62*, 849–857. [CrossRef]

23. Meijs, S.; Fjorback, M.; Jensen, C.; Sørensen, S.; Rechendorff, K.; Rijkhoff, N.J.M. Electrochemical properties of titanium nitride nerve stimulation electrodes: An in vitro and in vivo study. *Front. Neurosci.* **2015**, *9*, 268. [CrossRef] [PubMed]

24. Meijs, S.; Fjorback, M.; Jensen, C.; Sørensen, S.; Rechendorff, K.; Rijkhoff, N.J.M. Influence of fibrous encapsulation on electro-chemical properties of TiN electrodes. *Med. Eng. Phys.* **2016**, *38*, 468–476. [CrossRef] [PubMed]

25. Meijs, S.; Sørensen, C.; Sørensen, S.; Rechendorff, K.; Fjorback, M.; Rijkhoff, N.J.M. Influence of implantation on the electrochemical properties of smooth and porous TiN coatings for stimulation electrodes. *J. Neural Eng.* **2016**, *13*, 026011. [CrossRef] [PubMed]

26. Meijs, S.; Alcaide, M.; Sørensen, C.; McDonald, M.; Sørensen, S.; Rechendorff, K.; Gerhardt, A.; Nesladek, M.; Rijkhoff, N.J.; Pennisi, C.P. Biofouling resistance of boron-doped diamond neural stimulation electrodes is superior to titanium nitride electrodes in vivo. *J. Neural Eng.* **2016**, *13*, 056011. [CrossRef] [PubMed]

27. Black, B.J.; Kanneganti, A.; Joshi-Imre, A.; Rihani, R.; Chakraborty, B.; Abbott, J.; Pancrazio, J.J.; Cogan, S.F. Chronic recording and electrochemical performance of Utah microelectrode arrays implanted in rat motor cortex. *J. Neurophys.* **2018**, *120*, 2083–2090. [CrossRef] [PubMed]

28. Lempka, S.F.; Miocinovic, S.; Johnson, M.D.; Vitek, J.L.; McIntyre, C.C. In vivo impedance spectroscopy of deep brain stimulation electrodes. *J. Neural Eng.* **2009**, *6*, 046001. [CrossRef] [PubMed]

29. Cyster, L.A.; Grant, D.M.; Parker, K.G.; Parker, T.L. The effect of surface chemistry and structure of titanium nitride (TiN) films on primary hippocampal cells. *Biomol. Eng.* **2002**, *19*, 171–175. [CrossRef]

30. Cyster, L.A.; Parker, K.G.; Parker, T.L.; Grant, D.M. The effect of surface chemistry and nanotopography of titanium nitride (TiN) films on 3T3-L1 fibroblasts. *J. Biomed. Mater. Res. A* **2003**, *67*, 138–147. [CrossRef]

31. Cyster, L.A.; Parker, K.G.; Parker, T.L.; Grant, D.M. The effect of surface chemistry and nanotopography of titanium nitride (TiN) films on primary hippocampal neurones. *Biomaterials* **2004**, *25*, 97–107. [CrossRef]

32. Massiani, Y.; Medjahed, A.; Crousier, J.P.; Gravier, P.; Rebatel, I. Corrosion of sputtered titanium nitride films deposited on iron and stainless steel. In *Metallurgical Coatings and Materials Surface Modifications*; Hintermann, H.E., Spitz, J., Eds.; North-Holland: Amsterdam, The Netherlands, 1991; pp. 115–120.

33. Avasarala, B.; Haldar, P. Electrochemical oxidation behavior of titanium nitride based electrocatalysts under PEM fuel cell conditions. *Electrochim. Acta* **2010**, *55*, 9024–9034. [CrossRef]

34. Norlin, A.; Pan, J.; Leygraf, C. Investigation of Pt, Ti, TiN and nano-porous carbon electrodes for implantable cardioverter-defibrillator applications. *Electrochim. Acta* **2004**, *49*, 4011–4020. [CrossRef]

35. Merrill, D.R.; Bikson, M.; Jefferys, J.G. Electrical stimulation of excitable tissue: Design of efficacious and safe protocols. *J. Neurosci. Methods* **2005**, *141*, 171–198. [CrossRef]

36. Ragheb, T.; Geddes, L.A. Electrical properties of metallic electrodes. *Med. Biol. Eng. Comput.* **1990**, *28*, 182–186. [CrossRef]

37. Meng, L.J.; Santos, M.D. Characterization of titanium nitride films prepared by dc reactive magnetron sputtering at different nitrogen pressures. *Surf. Coat. Technol.* **1997**, *90*, 64–70. [CrossRef]

38. Chawla, V.; Jayaganthan, R.; Chandra, R. Structural characterizations of magnetron sputtered nanocrystalline TiN thin films. *Mater. Charact.* **2008**, *59*, 1015–1020. [CrossRef]

39. Cunha, L.T.; Pedrosa, P.; Tavares, C.J.; Alves, E.; Vaz, F.; Fonseca, C. The role of composition, morphology and crystalline structure in the electrochemical behaviour of TiNx thin films for dry electrode sensor materials. *Electrochim. Acta* **2009**, *55*, 59–67. [CrossRef]

40. Sánchez, G.; Rodrigo, A.; Bologna Alles, A. Titanium nitride pacing electrodes with high surface-to-area ratios. *Acta Mater.* **2005**, *53*, 4079. [CrossRef]

41. Brummer, S.B.; Turner, M.J. Electrical stimulation with Pt electrodes: II-estimation of maximum surface redox (theoretical non-gassing) limits. *IEEE Trans. Biomed. Eng.* **1977**, *5*, 440–443. [CrossRef]

42. McHardy, J.; Robblee, L.S.; Marston, J.M.; Brummer, S.B. Electrical stimulation with Pt electrodes. IV. Factors influencing Pt dissolution in inorganic saline. *Biomaterials* **1980**, *1*, 129–134. [CrossRef]

43. Negi, S.; Bhandari, R.; Rieth, L.; Van Wagenen, R.; Solzbacher, F. Neural electrode degradation from continuous electrical stimulation: Comparison of sputtered and activated iridium oxide. *J. Neurosci. Methods* **2010**, *186*, 8–17. [CrossRef]

44. Musa, S.; Rand, D.R.; Bartic, C.; Eberle, W.; Nuttin, B.; Borghs, G. Coulometric detection of irreversible electrochemical reactions occurring at Pt microelectrodes used for neural stimulation. *Anal. Chem.* **2011**, *83*, 4012–4022. [CrossRef]

45. Norlin, A.; Pan, J.; Leygraf, C. Investigation of interfacial capacitance of Pt, Ti and TiN coated electrodes by electrochemical impedance spectroscopy. *Biomol. Eng.* **2002**, *19*, 67–71. [CrossRef]

46. Robblee, L.S.; McHardy, J.; Marston, J.M.; Brummer, S.B. Electrical stimulation with Pt electrodes. V. The effect of protein on Pt dissolution. *Biomaterials* **1980**, *1*, 135–139. [CrossRef]

47. Robblee, L.S.; McHardy, J.; Agnew, W.F.; Bullara, L.A. Electrical stimulation with Pt electrodes. VII. Dissolution of Pt electrodes during electrical stimulation of the cat cerebral cortex. *J. Neurosci. Methods* **1983**, *9*, 301–308. [CrossRef]

48. Shepherd, R.K.; Murray, M.T.; Hougiton, M.E.; Clark, G.M. Scanning electron microscopy of chronically stimulated platinum intracochlear electrodes. *Biomaterials* **1985**, *6*, 237–242. [CrossRef]

49. Hibbert, D.B.; Weitzner, K.; Tabor, B.; Carter, P. Mass changes and dissolution of platinum during electrical stimulation in artificial perilymph solution. *Biomaterials* **2000**, *21*, 2177–2182. [CrossRef]

50. McCreery, D.B.; Agnew, W.F.; Yuen, T.G.H.; Bullara, L.A. Relationship between stimulus amplitude, stimulus frequency and neural damage during electrical stimulation of sciatic nerve of cat. *Med. Biol. Eng. Comput.* **1995**, *33*, 426–429. [CrossRef]

51. Cogan, S.F.; Ludwig, K.A.; Welle, C.G.; Takmakov, P. Tissue damage thresholds during therapeutic electrical stimulation. *J. Neural Eng.* **2016**, *13*, 021001. [CrossRef] [PubMed]

52. Meijs, S. The Influence of Tissue Responses on the Electrochemical Properties of Implanted Neural Stimulation Electrodes. Ph.D. Thesis, Aalborg Universitetsforlag, Aalborg, Denmark, 20 May 2016.

53. Loeb, G.E.; Richmond, F.J.; Baker, L.L. The BION devices: Injectable interfaces with peripheral nerves and muscles. *Neurosurg. Focus* **2006**, *20*, 1–9. [CrossRef]

54. Majid, A. *Electroceuticals*, 1st ed.; Springer: Basel, Switzerland, 2017.

55. Datta, S.; Das, M.; Balla, V.K.; Bodhak, S.; Murugesan, V.K. Mechanical, wear, corrosion and biological properties of arc deposited titanium nitride coatings. *Surf. Coat. Technol.* **2018**, *344*, 214–222. [CrossRef]

MDPI

St. Alban-Anlage 66

4052 Basel

Switzerland

Tel. +41 61 683 77 34

Fax +41 61 302 89 18

www.mdpi.com

Metals Editorial Office

E-mail: metals@mdpi.com

www.mdpi.com/journal/metals